T0292633

Studies in Computational Intelligence

Volume 629

About this Series

The series "Studies in Computational Intelligence" (SCI) publishes new developments and advances in the various areas of computational intelligence—quickly and with a high quality. The intent is to cover the theory, applications, and design methods of computational intelligence, as embedded in the fields of engineering, computer science, physics and life sciences, as well as the methodologies behind them. The series contains monographs, lecture notes and edited volumes in computational intelligence spanning the areas of neural networks, connectionist systems, genetic algorithms, evolutionary computation, artificial intelligence, cellular automata, self-organizing systems, soft computing, fuzzy systems, and hybrid intelligent systems. Of particular value to both the contributors and the readership are the short publication timeframe and the worldwide distribution, which enable both wide and rapid dissemination of research output.

More information about this series at http://www.springer.com/series/7092

Erik Cuevas · Margarita Arimatea Díaz Cortés
Diego Alberto Oliva Navarro

Advances of Evolutionary Computation: Methods and Operators

Erik Cuevas
Departamento de Electrónica, CUCEI
Universidad de Guadalajara
Guadalajara, Jalisco
Mexico

Margarita Arimatea Díaz Cortés
Departamento de Electrónica, CUCEI
Universidad de Guadalajara
Guadalajara, Jalisco
Mexico

and

Institut für Informatik
Freie Universität Berlin
Berlin
Germany

Diego Alberto Oliva Navarro
Departamento de Electrónica, CUCEI
Universidad de Guadalajara
Guadalajara, Jalisco
Mexico

and

Tecnológico de Monterrey
Campus Guadalajara
Zapopan
Mexico

ISSN 1860-949X ISSN 1860-9503 (electronic)
Studies in Computational Intelligence
ISBN 978-3-319-28502-3 ISBN 978-3-319-28503-0 (eBook)
DOI 10.1007/978-3-319-28503-0

Library of Congress Control Number: 2015960768

Printed on acid-free paper

This Springer imprint is published by SpringerNature
The registered company is Springer International Publishing AG Switzerland

To my beloved mother, always my support

Margarita Arimatea Díaz Cortés

Preface

Evolutionary computation (EC) is one of the most important emerging technologies of recent times. Over the last years, there has been exponential growth of research activity in this field. Despite the fact that the concept itself has not been precisely defined, EC has become the standard term that encompasses several stochastic, population-based, and system-inspired approaches.

EC methods use as inspiration our scientific understanding of biological, natural, or social systems, which at some level of abstraction can be represented as optimization processes. They intend to serve as general-purpose easy-to-use optimization techniques capable of reaching globally optimal or at least nearly optimal solutions. In their operation, searcher agents emulate a group of biological or social entities which interact to each other based on specialized operators that model a determined biological or social behavior. These operators are applied to a population (or several subpopulations) of candidate solutions (individuals) that are evaluated with respect to their fitness. Thus, in the evolutionary process, individual positions are successively approximated to the optimal solution of the system to be solved.

Due to their robustness, EC techniques are well-suited options for industrial and real-world tasks. They do not need gradient information, and they can operate on each kind of parameter space (continuous, discrete, combinatorial, or even mixed variants). Essentially, the credibility of evolutionary algorithms relies on their ability to solve difficult, real-world problems with the minimal amount of human effort.

There exist some common features clearly appear in most of the EC approaches, such as the use of diversification to force the exploration of regions of the search space, rarely visited until now, and the use of intensification or exploitation, to investigate thoroughly some promising regions. Another common feature is the use of memory to archive the best solutions encountered.

Numerous books have been published tacking into account any of the most widely known methods, namely simulated annealing, tabu search, evolutionary algorithms, ant colony algorithms, particle swarm optimization, or differential evolution, but attempts to consider the discussion of alternative approaches are scarce.

The excessive publication of developments based on the simple modification of popular EC methods presents an important disadvantage, in that it distracts attention away from other innovative ideas in the field of EC. There exist several alternative EC methods which consider very interesting concepts; however, they seem to have been completely overlooked in favor of the idea of modifying, hybridizing, or restructuring traditional EC approaches.

The goal of this book is to present advances that discuss alternative EC developments or highlight non-conventional operators which prove to be effective in adapting a determined EC method to a specific problem. This book has been structured so that each chapter can be read independently from the others. Chapter 1 describes evolutionary computation (EC). This chapter concentrates on elementary concepts of evolutionary algorithms. Readers that are familiar with EC may wish to skip this chapter.

In Chap. 2, a swarm algorithm, namely the Social Spider Optimization (SSO), is presented for solving optimization tasks. The SSO algorithm is based on the simulation of the cooperative behavior of social spiders. In SSO, individuals emulate a group of spiders which interact to each other based on the biological laws of the cooperative colony. Different to the most EC algorithms, SSO considers two different search agents (spiders): males and females. Depending on the gender, each individual is conducted by a set of different evolutionary operators which mimic the different cooperative behaviors assumed in the colony. To illustrate the proficiency and robustness of the SSO, it is compared to other well-known evolutionary methods.

Chapter 3 presents a nature-inspired algorithm called the States of Matter Search (SMS). The SMS algorithm is based on the modeling of the states of matter phenomenon. In SMS, individuals emulate molecules which interact to each other by using evolutionary operations based on the physical principles of the thermal energy motion mechanism. The algorithm is devised considering each state of matter one different exploration–exploitation ratio. In SMS, the evolutionary process is divided into three phases which emulate the three states of matter: gas, liquid, and solid. In each state, the evolving elements exhibit different movement capacities. Beginning from the gas state (pure exploration), the algorithm modifies the intensities of exploration and exploitation until the solid state (pure exploitation) is reached. As a result, the approach can substantially improve the balance between exploration–exploitation, yet preserving the good search capabilities of an EC method. To illustrate the proficiency and robustness of the proposed algorithm, it was compared with other well-known evolutionary methods including recent variants that incorporate diversity preservation schemas.

In Chap. 4, an EC algorithm inspired by the collective animal behavior (CAB) is presented. In this algorithm, the searcher agents represent a group of animals that interact to each other based on simple behavioral rules which are modeled as mathematical operators. Such operations are applied to each agent considering that the complete group has a memory storing their own best positions seen so far, by using a simple competition principle. The approach has been compared to other well-known optimization methods. The results confirm a high performance of the proposed method for solving various benchmark functions.

In Chap. 5, a novel biologically inspired algorithm, namely Allostatic Optimization (AO), is presented. The AO algorithm uses as metaphor the allostasis mechanisms for designing an optimization methodology. AO provides a population-based search procedure under which all individuals, also seen as body conditions, are defined within the multidimensional space. They are generated or modified by using several evolutionary operators that emulate the different operations employed by the allostasis process, whereas the fitness function evaluates the capacity of each individual (body condition) to reach a steady health state (good solution). Different to several popular EC methods, AO implements evolutionary operators that avoid concentrating most of the particles in only one position favoring the exploration process and eliminating the flaws related to the premature convergence. The approach has been compared to other well-known evolutionary algorithms. The results confirm a high performance of the proposed method for solving various benchmark functions.

In Chap. 6, a swarm algorithm, called the Locust Search (LS), is presented for solving some optimization tasks. The LS algorithm is based on the behavioral modeling of swarms of locusts. In LS, individuals represent a group of locusts which interact to each other based on the biological laws of the cooperative swarm. The algorithm considers two different behaviors: solitary and social. Depending on the behavior, each individual is conducted by a set of evolutionary operators which mimics different cooperative conducts that are typically found in the swarm. Different to most of existent swarm algorithms, the behavioral model in the proposed approach explicitly avoids the concentration of individuals in the current best positions. Such fact allows not only to emulate in a better realistic way the cooperative behavior of the locust colony, but also to incorporate a computational mechanism to avoid critical flaws that are commonly present in the popular particle swarm optimization and differential evolution, such as the premature convergence and the incorrect exploration–exploitation balance. In order to illustrate the proficiency and robustness of the proposed approach, its performance is compared to other well-known evolutionary methods. The comparison examines several standard benchmark functions which are commonly considered in the literature.

Chapter 7 presents an EC method called the Adaptive Population with Reduced Evaluations (APREs) for solving optimization problems which are characterized by demanding an excessive number of function evaluations. APRE reduces the number of function evaluations through the use of two mechanisms: (1) adapting dynamically the size of the population and (2) incorporating a fitness estimation strategy that decides the amount of individuals to be evaluated with the original fitness function and the amount of individuals to be estimated by a very simple approximated model. APRE begins with an initial random population which will be used as a memory during the evolution process. After initialization, it is selected the elements to be evolved. Its number is automatically modified in each iteration. With the selected elements, a set of new individuals is generated as a consequence of the execution of the seeking operation. Afterward, the memory is updated. For this process, the new individuals produced by the seeking operation compete against the memory elements to build the final memory configuration. Finally, a sample of the best elements contained in the final memory configuration is undergone to the refinement operation. This cycle is repeated until the maximum number the iterations have been reached. Different to other approaches that use an already existent EA as framework, the APRE method has been completely designed to substantially reduce the computational cost without degrading its good search capacities.

Different to global optimization, the main objective of multimodal optimization is to find multiple global and local optima for a problem in one single run. Finding multiple solutions to a multimodal optimization problem is especially useful in engineering, since the best solution may not always be the best realizable due to various practical constraints. In Chap. 8, the multimodal characteristics of the CAB algorithm are exposed in Chap. 4. The main objective is to analyze the particular CAB operators that permit its multimodal performance.

Finally, Chap. 9 presents a variant of the SSO algorithm (exposed in Chap. 2) for solving constrained optimization problems. The method, called SSO-C, implements additional mechanisms that the original one. For constraint handling, SSO-C incorporates the combination of two different paradigms in order to direct the search toward feasible regions of the search space. In particular, it has been added: (1) a penalty function which introduces a tendency term into the original objective function to penalize constraint violations in order to solve a constrained problem as an unconstrained one and (2) a feasibility criterion to bias the generation of new individuals toward feasible regions increasing also their probability of getting better solutions.

As authors, we wish to thank many people who were somehow involved in the writing process of this book. We thank Dr. Raul Rojas and Dr. Gonzalo Pajares for supporting us to have it published. We express our gratitude to Prof. Janusz kacprzyk, who so warmly sustained this project. Acknowledgements also go to Dr. Thomas Ditzinger, who so kindly agreed to its appearance. We also acknowledge that this work was supported by CONACYT under the grant number CB 181053.

Guadalajara
Berlin
Zapopan
December 2015

Erik Cuevas
Margarita Arimatea Díaz Cortés
Diego Alberto Oliva Navarro

Contents

Chapter 1
Introduction

1.1 Definition of an Optimization Problem

The vast majority of image processing and pattern recognition algorithms use some form of optimization, as they intend to find some solution which is "best" according to some criterion. From a general perspective, an optimization problem is a situation that requires to decide for a choice from a set of possible alternatives in order to reach a predefined/required benefit at minimal costs [1].

Consider a public transportation system of a city, for example. Here the system has to find the "best" route to a destination location. In order to rate alternative solutions and eventually find out which solution is "best," a suitable criterion has to be applied. A reasonable criterion could be the distance of the routes. We then would expect the optimization algorithm to select the route of shortest distance as a solution. Observe, however, that other criteria are possible, which might lead to different "optimal" solutions, e.g., number of transfers, ticket price or the time it takes to travel the route leading to the fastest route as a solution.

Mathematically speaking, optimization can be described as follows: Given a function $f : S \to \mathbb{R}$ which is called the objective function, find the argument which minimizes f:

$$x^* = \arg\min_{x \in S} f(x) \tag{1.1}$$

S defines the so-called solution set, which is the set of all possible solutions for the optimization problem. Sometimes, the unknown(s) x are referred to design variables. The function f describes the optimization criterion, i.e., enables us to calculate a quantity which indicates the "quality" of a particular x.

E. Cuevas et al., *Advances of Evolutionary Computation: Methods and Operators*, Studies in Computational Intelligence 629,
DOI 10.1007/978-3-319-28503-0_1

In our example, S is composed by the subway trajectories and bus lines, etc., stored in the database of the system, x is the route the system has to find, and the optimization criterion $f(x)$ (which measures the quality of a possible solution) could calculate the ticket price or distance to the destination (or a combination of both), depending on our preferences.

Sometimes there also exist one or more additional constraints which the solution x^* has to satisfy. In that case we talk about constrained optimization (opposed to unconstrained optimization if no such constraint exists). As a summary, an optimization problem has the following components:

- One or more design variables x for which a solution has to be found
- An objective function $f(x)$ describing the optimization criterion
- A solution set S specifying the set of possible solutions x
- (optional) One or more constraints on x

In order to be of practical use, an optimization algorithm has to find a solution in a reasonable amount of time with reasonable accuracy. Apart from the performance of the algorithm employed, this also depends on the problem at hand itself. If we can hope for a numerical solution, we say that the problem is well-posed. For assessing whether an optimization problem is well-posed, the following conditions must be fulfilled:

1. A solution exists.
2. There is only one solution to the problem, i.e., the solution is unique.
3. The relationship between the solution and the initial conditions is such that small perturbations of the initial conditions result in only small variations of x^*.

1.2 Classical Optimization

Once a task has been transformed into an objective function minimization problem, the next step is to choose an appropriate optimizer. Optimization algorithms can be divided in two groups: derivative-based and derivative-free [2].

In general, $f(x)$ may have a nonlinear form respect to the adjustable parameter x. Due to the complexity of $f(\cdot)$, in classical methods, it is often used an iterative algorithm to explore the input space effectively. In iterative descent methods, the next point x_{k+1} is determined by a step down from the current point x_k in a direction vector $\mathbf{d}$:

$$x_{k+1} = x_k + \alpha\mathbf{d}, \tag{1.2}$$

where α is a positive step size regulating to what extent to proceed in that direction. When the direction d in Eq. 1.1 is determined on the basis of the gradient ($\mathbf{g}$) of the objective function $f(\cdot)$, such methods are known as gradient-based techniques.

The method of steepest descent is one of the oldest techniques for optimizing a given function. This technique represents the basis for many derivative-based methods. Under such a method, the Eq. 1.3 becomes the well-known gradient formula:

$$x_{k+1} = x_k - \alpha \mathbf{g}(f(x)), \tag{1.3}$$

However, classical derivative-based optimization can be effective as long the objective function fulfills two requirements:

- The objective function must be two-times differentiable.
- The objective function must be uni-modal, i.e., have a single minimum.

A simple example of a differentiable and uni-modal objective function is

$$f(x_1, x_2) = 10 - \mathrm{e}^{-\left(x_1^2 + 3 \cdot x_2^2\right)} \tag{1.4}$$

Figure 1.1 shows the function defined in Eq. 1.4.

Unfortunately, under such circumstances, classical methods are only applicable for a few types of optimization problems. For combinatorial optimization, there is no definition of differentiation.

Furthermore, there are many reasons why an objective function might not be differentiable. For example, the “floor” operation in Eq. 1.5 quantizes the function in Eq. 1.4, transforming Fig. 1.1 into the stepped shape seen in Fig. 1.2. At each step’s edge, the objective function is non-differentiable:

$$f(x_1, x_2) = \mathrm{floor}\left(10 - \mathrm{e}^{-\left(x_1^2 + 3 \cdot x_2^2\right)}\right) \tag{1.5}$$

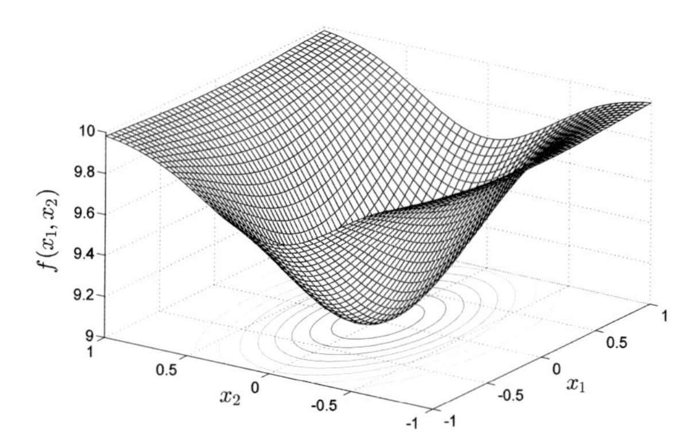

Fig. 1.1 Uni-modal objective function

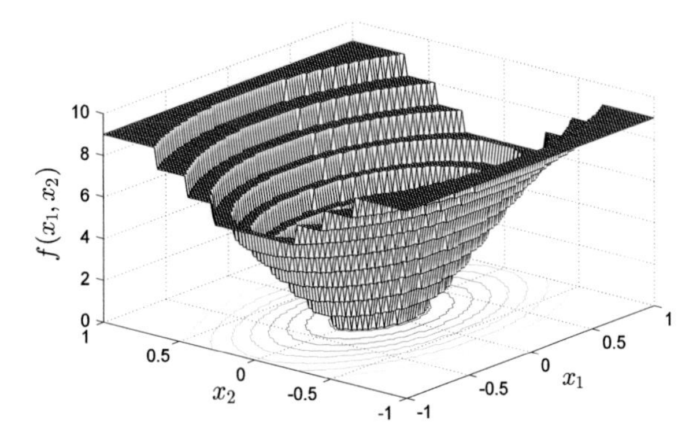

Fig. 1.2 A non-differentiable, quantized, uni-modal function

Even in differentiable objective functions, gradient-based methods might not work. Let us consider the minimization of the Griewank function as an example.

$$\begin{array}{ll} \text{minimize} & f(x_1, x_2) = \dfrac{x_1^2 + x_2^2}{4000} - \cos(x_1)\cos\left(\dfrac{x_2}{\sqrt{2}}\right) + 1 \\ \text{subject to} & \begin{array}{c} -30 \le x_1 \le 30 \\ -30 \le x_2 \le 30 \end{array} \end{array} \tag{1.6}$$

From the optimization problem formulated in Eq. 1.6, it is quite easy to understand that the global optimal solution is $x_1 = x_2 = 0$. Figure 1.3 visualizes the function defined in Eq. 1.6. According to Fig. 1.3, the objective function has many local optimal solutions (multimodal) so that the gradient methods with a randomly generated initial solution will converge to one of them with a large probability.

Considering the limitations of gradient-based methods, image processing and pattern recognition problems make difficult their integration with classical optimization methods. Instead, some other techniques which do not make assumptions and which can be applied to wide range of problems are required [3].

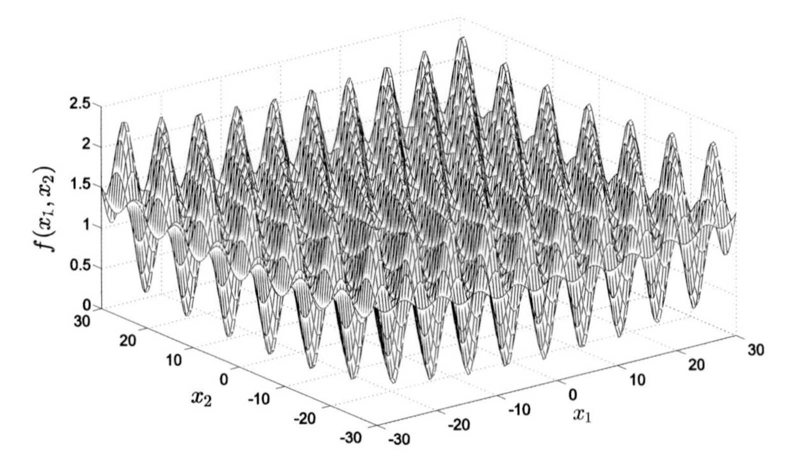

Fig. 1.3 The Griewank multi-modal function

1.3 Evolutionary Computation Methods

Evolutionary computation (EC) [4] methods are derivative-free procedures, which do not require that the objective function must be neither two-times differentiable nor uni-modal. Therefore, EC methods as global optimization algorithms can deal with non-convex, nonlinear, and multimodal problems subject to linear or nonlinear constraints with continuous or discrete decision variables.

The field of EC has a rich history. With the development of computational devices and demands of industrial processes, the necessity to solve some optimization problems arose despite the fact that there was not sufficient prior knowledge (hypotheses) on the optimization problem for the application of a classical method. In fact, in the majority of image processing and pattern recognition cases, the problems are highly nonlinear, or characterized by a noisy fitness, or without an explicit analytical expression as the objective function might be the result of an experimental or simulation process. In this context, the EC methods have been proposed as optimization alternatives.

An EC technique is a general method for solving optimization problems. It uses an objective function in an abstract and efficient manner, typically without utilizing deeper insights into its mathematical properties. EC methods do not require hypotheses on the optimization problem nor any kind of prior knowledge on the objective function. The treatment of objective functions as "black boxes" [5] is the most prominent and attractive feature of EC methods.

EC methods obtain knowledge about the structure of an optimization problem by utilizing information obtained from the possible solutions (i.e., candidate solutions) evaluated in the past. This knowledge is used to construct new candidate solutions which are likely to have a better quality.

Recently, several EC methods have been proposed with interesting results. Such approaches uses as inspiration our scientific understanding of biological, natural or social systems, which at some level of abstraction can be represented as optimization processes [6]. These methods include the social behavior of bird flocking and fish schooling such as the Particle Swarm Optimization (PSO) algorithm [7], the cooperative behavior of bee colonies such as the Artificial Bee Colony (ABC) technique [8], the improvisation process that occurs when a musician searches for a better state of harmony such as the Harmony Search (HS) [9], the emulation of the bat behavior such as the Bat Algorithm (BA) method [10], the mating behavior of firefly insects such as the Firefly (FF) method [11], the social-spider behavior such as the Social Spider Optimization (SSO) [12], the simulation of the animal behavior in a group such as the Collective Animal Behavior [13], the emulation of immunological systems as the clonal selection algorithm (CSA) [14], the simulation of the electromagnetism phenomenon as the electromagnetism-Like algorithm [15], and the emulation of the differential and conventional evolution in species such as the Differential Evolution (DE) [16] and Genetic Algorithms (GA) [17], respectively.

1.3.1 Structure of an Evolutionary Computation Algorithm

From a conventional point of view, an EC method is an algorithm that simulates at some level of abstraction a biological, natural or social system. To be more specific, a standard EC algorithm includes:

1. One or more populations of candidate solutions are considered.
2. These populations change dynamically due to the production of new solutions.
3. A fitness function reflects the ability of a solution to survive and reproduce.
4. Several operators are employed in order to explore an exploit appropriately the space of solutions.

The EC methodology suggests that, on average, candidate solutions improve their fitness over generations (i.e., their capability of solving the optimization problem). A simulation of the evolution process based on a set of candidate solutions whose fitness is properly correlated to the objective function to optimize will, on average, lead to an improvement of their fitness and thus steer the simulated population towards the global solution.

Most of the optimization methods have been designed to solve the problem of finding a global solution of a nonlinear optimization problem with box constraints in the following form:

$$\begin{array}{lll} \text{maximize} & f(\mathbf{x}), & \mathbf{x} = (x_1, \ldots, x_d) \in \mathbb{R}^d \\ \text{subject to} & \mathbf{x} \in \mathbf{X} & \end{array} \tag{1.7}$$

where $f : \mathbb{R}^d \to \mathbb{R}$ is a nonlinear function whereas $\mathbf{X} = \{\mathbf{x} \in \mathbb{R}^d | l_i \le x_i \le u_i,\ i = 1, \ldots, d.\}$ is a bounded feasible search space, constrained by the lower (l_i) and upper (u_i) limits.

In order to solve the problem formulated in Eq. 1.6, in an evolutionary computation method, a population $\mathbf{P}^k(\{\mathbf{p}_1^k, \mathbf{p}_2^k, \ldots, \mathbf{p}_N^k\})$ of N candidate solutions (individuals) evolves from the initial point (k = 0) to a total *gen* number iterations (k = *gen*). In its initial point, the algorithm begins by initializing the set of N candidate solutions with values that are randomly and uniformly distributed between the pre-specified lower (l_i) and upper (u_i) limits. In each iteration, a set of evolutionary operators are applied over the population $\mathbf{P}^k$ to build the new population $\mathbf{P}^{k+1}$. Each candidate solution $\mathbf{p}_i^k (i \in [1, \ldots, N])$ represents a d-dimensional vector $\left\{p_{i,1}^k, p_{i,2}^k, \ldots, p_{i,d}^k\right\}$ where each dimension corresponds to a decision variable of the optimization problem at hand. The quality of each candidate solution $\mathbf{p}_i^k$ is evaluated by using an objective function $f\left(\mathbf{p}_i^k\right)$ whose final result represents the fitness value of $\mathbf{p}_i^k$. During the evolution process, the best candidate solution $\mathbf{g}$ $(g_1, g_2, \ldots g_d)$ seen so-far is preserved considering that it represents the best available solution. Figure 1.4 presents a graphical representation of a basic cycle of a EC method.

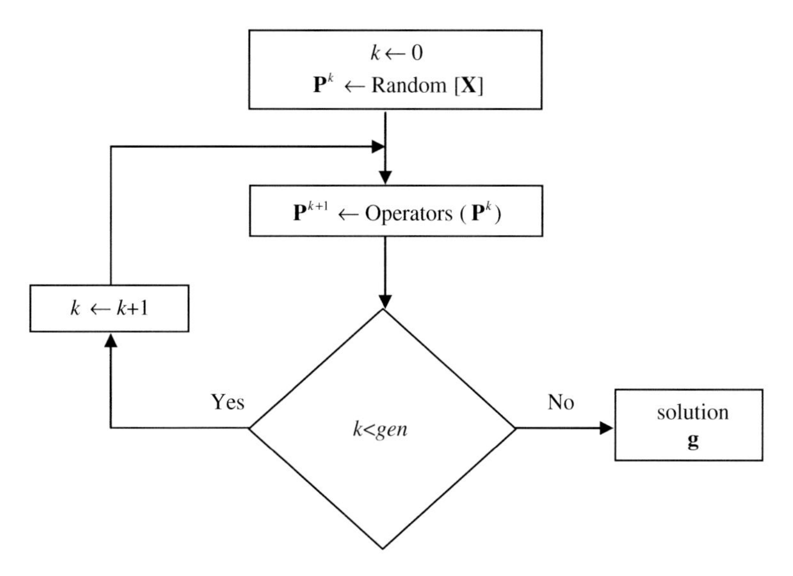

Fig. 1.4 The basic cycle of a EC method

References

1. Akay, B., Karaboga, D.: A survey on the applications of artificial bee colony in signal, image, and video processing. SIViP **9**(4), 967–990 (2015)
2. Yang, X.-S.: *Engineering Optimization*. Wiley, New York (2010)
3. Alexander Treiber, M.: Optimization for Computer Vision an Introduction to Core Concepts and Methods. Springer, New York (2013)
4. Simon, D.: Evolutionary Optimization Algorithms. Wiley, New York (2013)
5. Blum, C., Roli, A.: Metaheuristics in combinatorial optimization: overview and conceptual comparison. ACM Comput. Surv. (CSUR) **35**(3), 268–308 (2003). doi:10.1145/937503.937505
6. Nanda, S.J., Panda, G.: A survey on nature inspired metaheuristic algorithms for partitional clustering. Swarm Evol. Comput. **16**, 1–18 (2014)
7. Kennedy, J., Eberhart, R.: Particle swarm optimization. In: Proceedings of the 1995 IEEE International Conference on Neural Networks, vol. 4, pp. 1942–1948 (1995)
8. Karaboga, D.: An idea based on honey bee swarm for numerical optimization. Technical Report-TR06. Engineering Faculty, Computer Engineering Department, Erciyes University (2005)
9. Geem, Z.W., Kim, J.H., Loganathan, G.V.: A new heuristic optimization algorithm: harmony search. Simulations **76**, 60–68 (2001)
10. Yang, X.S.: A new metaheuristic bat-inspired algorithm. In: Cruz, C., González, J., Krasnogor, G.T.N., Pelta, D.A. (eds.) Nature Inspired Cooperative Strategies for Optimization (NISCO 2010), Studies in Computational Intelligence, vol. 284, pp. 65–74. Springer, Berlin (2010)
11. Yang, X.S.: Firefly algorithms for multimodal optimization. In: Stochastic Algorithms: Foundations and Applications, SAGA 2009, vol. 5792, pp. 169–178. Lecture Notes in Computer Sciences (2009)
12. Cuevas, E., Cienfuegos, M., Zaldívar, D., Pérez-Cisneros, M.: A swarm optimization algorithm inspired in the behavior of the social-spider. Expert Syst. Appl. **40**(16), 6374–6384 (2013)

13. Cuevas, E., González, M., Zaldivar, D., Pérez-Cisneros, M., García, G.: An algorithm for global optimization inspired by collective animal behaviour. Discrete Dyn. Nat. Soc., art. no. 638275 (2012)
14. de Castro, L.N., von Zuben, F.J.: Learning and optimization using the clonal selection principle. IEEE Trans. Evol. Comput. **6**(3), 239–251 (2002)
15. Birbil, Ş.I., Fang, S.C.: An electromagnetism-like mechanism for global optimization. J. Glob. Optim. **25**(1), 263–282 (2003)
16. Storn, R., Price, K.: Differential evolution—A simple and efficient adaptive scheme for global optimisation over continuous spaces. Technical Report TR-95–012. ICSI, Berkeley, CA (1995)
17. Goldberg, D.E.: Genetic Algorithm in Search Optimization and Machine Learning. Addison-Wesley, Reading (1989)

Chapter 2
A Swarm Global Optimization Algorithm Inspired in the Behavior of the Social-Spider

2.1 Introduction

The collective intelligent behavior of insect or animal groups in nature such as flocks of birds, colonies of ants, schools of fish, swarms of bees, and termites have attracted the attention of researchers. The aggregate behavior of insects or animals is called swarm behavior. Entomologists have studied this collective phenomenon to model biological swarms, and engineers applied these models as a framework for solving complex real-world problems. This branch of artificial intelligence which deals with the collective behavior of swarms through complex interaction of individuals without supervision is referred to as swarm intelligence. Bonabeau defined swarm intelligence as "any attempt to design algorithms or distributed problem solving devices inspired by the collective behavior of the social insect colonies and other animal societies" [1]. Swarm intelligence has some advantages such as scalability, fault tolerance, adaptation, speed, modularity, autonomy, and parallelism [2].

The key components of swarm intelligence are self-organization and division of labors. In a self-organizing system, each of the covered units may respond to local stimuli individually and act together to accomplish a global task via division of labors without a centralized supervision. The entire system can adapt to internal and external changes efficiently.

Several swarm algorithms have been developed by a combination of deterministic rules and randomness, mimicking the behavior of insect or animal groups in nature. Such methods include the social behavior of bird flocking and fish schooling such as the Particle Swarm Optimization (PSO) algorithm [3], the cooperative behavior of bee colonies such as the Artificial Bee Colony (ABC) technique [4], the social foraging behavior of bacteria such as the Bacterial Foraging Optimization Algorithm (BFOA) [5], the simulation of the herding behavior of krill individuals such as the Krill Herd (KH) method [6], the mating behavior of firefly insects such as the Firefly (FF) method [7] and the emulation of

E. Cuevas et al., *Advances of Evolutionary Computation: Methods and Operators*, Studies in Computational Intelligence 629,
DOI 10.1007/978-3-319-28503-0_2

the lifestyle of cuckoo birds such as the Cuckoo Optimization Algorithm (COA) [8].

Insect colonies and animal groups provide a rich set of metaphors for designing swarm optimization algorithms. Such cooperative entities are complex system composed by individuals with different cooperative-tasks where each member tends to reproduce specialized behaviors depending generally on its gender [9]. However, most of the swarm algorithms model individuals as unisex performing virtually the same behavior. Under these circumstances, algorithms waste the possibility to add new and selective operators as a result of considering individuals with different characteristics such as sex, task-responsibility, etc. These operators could incorporate computational mechanisms to improve several important algorithm characteristics such as population diversity or searching capacities.

Although PSO and ABC are the most popular swarm algorithms for solving complex optimization problems, they present serious flaws such as premature convergence and difficulty to overcome local minima [10, 11]. The reason of these problems is the operators used for modifying the individual positions. In such algorithms, during their evolution, the position of each agent in the next iteration is updated yielding an attraction towards the position of the best particle seen so-far (in case of PSO) or of other randomly chosen individual (in case of ABC). Such behaviors produce that the entire population, as the algorithm evolves, concentrates around the best particle or diverges without control, favoring the premature convergence or damaging the exploration-exploitation balance [12, 13].

The interesting and exotic collective behaviors of social insects have fascinated and attracted the interest of researchers for many years. The collaborative swarming behavior that we observe in these groups provides survival advantages, where insect aggregations of relatively simple and "unintelligent" individuals can accomplish very complex tasks using only limited local information and simple rules of behavior [14]. Social-spiders are a representative example of social insects [15]. A social-spider is a spider species whose members maintain a set of complex cooperative behaviors [16]. Whereas most spiders are solitary and even aggressive toward other members of their own species, social-spiders show a tendency to live in groups, forming long-lasting aggregations, often referred to as colonies [17]. In a social-spider colony, each member, depending on its sex, executes a variety of tasks, such as predation, mating, web design, and social interaction [17, 18]. The web, as an important part of the colony, is not only used as a common environment for all members, but also as a communication channel among them [19] Therefore, important information (such as trapped prays or mating possibilities) is transmitted through the web in form of small vibrations. Such information, considered as a local knowledge, is employed by each member to conduct its own cooperative behavior, influencing simultaneously the social regulation of the colony [20].

In this chapter, a novel swarm algorithm, namely the Social Spider Optimization (SSO) is presented for solving optimization tasks. The SSO algorithm is based on the simulation of the cooperative behavior of social-spiders. In the presented algorithm, individuals emulate a group of spiders which interact to each other based on the biological laws of the cooperative colony. The algorithm considers two different search agents (spiders): males and females. Depending on the sex, each individual is conducted by a set of different evolutionary operators which mimic the different cooperative behaviors assumed in the colony. Different to most of the existent swarm algorithms, in the presented approach, each individual is modeled considered two different genders. Such fact allows not only to emulate in a better realistic way the cooperative behavior of the colony, but also to incorporate computational mechanisms to avoid critical flaws present in the popular PSO and ABC algorithms, such as premature convergence and incorrect exploration-exploitation balance. To illustrate the proficiency and robustness of the presented approach, it is compared to other well-known evolutionary methods. The comparison examines several standard benchmark functions which are commonly considered within the literature of evolutionary algorithms. The results show a high performance of the presented method when searching for a global optimum of several benchmark functions.

This chapter is organized as follows. In Sect. 2.2, are introduced the basic biological aspects of the algorithm. In Sect. 2.3, the novel SSO algorithm and its characteristics are both described. Section 2.4 presents the experimental results and the comparative study. Finally, in Sect. 2.5, conclusions are drawn.

2.2 Biologic Fundamentals

Social insect societies are complex cooperative systems that self-organize within a set of constraints. Cooperative groups are better at manipulating and exploiting their environment, defending resources and brood, and allow for task specialization among group members [21, 22]. A social insect colony functions as an integrated unit that not only possesses the ability to operate in a distributed manner, but also undertake enormous construction of global projects [23]. It is important to acknowledge that global order in social insects can arise as a result of internal interactions among insects.

A few species of spiders have been documented exhibiting a degree of social behavior [15]. One can generalize the behavior of these species in two basic forms, solitary spiders and social spiders [17]. This classification is made based on the level of cooperative behavior that they exhibit [18]. In general, solitary spiders create and maintain their own web while live in scarce contact with other individuals of the same species. In contrast, social spiders form colonies that remain together on a communal web, where a close spatial separation is presented between group members [19].

A social spider colony is composed of two fundamental components: members and a communal web. Members are divided in two different categories, males and females. An interesting characteristic of social-spiders is the highly female-biased populations. Some studies suggest that the number of male spiders barely reaches the 30 % of the total colony members [17, 24]. In the colony, each member, depending on its gender, cooperate in different activities such as build and maintain the communal web, prey capture, mating and social contact [20]. Interactions among members are either direct or indirect [25]. Direct interactions imply body contact, or the exchange of fluid, such as mating. For indirect interactions, it is used the communal web as a "medium of communication". Through the communal web, it is transmitted important information available for each colony member [19]. This information, encoded in form of small vibrations, is a critical aspect for the collective coordination among the members [20]. Since the vibrations depend on the weight and distance of the elements which provoke them, they are employed by the colony members to decode several messages, such as size of the trapped preys, characteristics of the neighboring members, etc.

In spite of the complexity, all the cooperative global patterns, presented in a colony level, are generated as a result of internal interactions among colony members [26]. Such internal iterations involve a set of simple behavioral rules followed by each spider in the colony. Behavioral rules are divided in two different classes: social interaction (cooperative behavior) and mating [27].

As a social insect, spiders perform cooperative interaction over other colony members. The way in which this behavior takes place depends on the spider gender. Female spiders which show a major tendency to socialize present an attraction or dislike over other spiders irrespective of the gender [17]. For a particular female spider, such attraction or dislike is commonly developed over other spiders that according to their vibrations (emitted over the communal web) represent strong colony members [20]. Since the vibrations depend on the weight and distance of the members which provoke them, strong vibrations are produced either by big spiders or neighboring members [19]. The bigger a spider is, the better it is considered as a colony member. The final decision of attraction or dislike over a determined member is taken according to an internal state which is influenced by several factors such as reproduction cycle, curiosity, and other random phenomena [20].

Different to female spiders, the behavior of male members is reproductive oriented [28]. Male spiders recognize themselves a subgroup of alpha males which dominate the colony resources. Therefore, the male population is divided in two classes: dominant and non-dominant male spiders [28]. Dominant male spiders have better fitness characteristics (normally size) in comparison with non-dominant. As a main behavior, dominant males are attracted to the closest female spider in the communal web. In contrast, non-dominant male spiders tend to concentrate in the center of the male population, as a strategy to take advantage of the resources wasted by the dominant males [29].

Mating is an important operation that no only assures the colony survival, but also allows the information exchange among members. Mating in a social-spider colony is performed by dominant males and the female members [30]. Under such circumstances, when a dominant male spider locates to one or more female members within a specific range, it mates with all the females in order to produce offspring [31].

2.3 The Social Spider Optimization (SSO) Algorithm

In this chapter, the operational principles from the social-spider colony have been used as guidelines for developing a new swarm optimization algorithm. The SSO assumes that entire search space is a communal web, where all the social-spiders interact. In the presented approach, each solution within the search space represents a spider position in the communal web. Every spider receives a weight according to the fitness value of the solution that the social-spider symbolizes. The algorithm models two different search agents (spiders): males and females. Depending on the gender, each individual is conducted by a set of different evolutionary operators which mimic the different cooperative behaviors assumed in the colony.

An interesting characteristic of social-spiders is the highly female-biased populations. In order to emulate this fact, the algorithm starts by defining the number of female and male spiders that will be characterized as individuals in the search space. The number of females N_f is randomly selected within the range of 65–90 % of the entire population N, previously chosen. Therefore, N_f is calculated by the following equation:

$$N_f = \text{floor}[(0.9 - \text{rand} \cdot 0.25) \cdot N] \tag{2.1}$$

where rand is a random number between [0, 1] whereas floor$(\cdot)$ maps a real number to an integer number. The number of male spiders N_m is computed as the complement between N and N_f. It is calculated as follows:

$$N_m = N - N_f \tag{2.2}$$

Therefore, the complete population S, composed by N elements, is divided in two sub-groups F and M. The Group F assembles the set of female individuals ($\mathbf{F} = \{\mathbf{f}_1, \mathbf{f}_2, \ldots, \mathbf{f}_{N_f}\}$) whereas **M** groups the male members ($\mathbf{M} = \{\mathbf{m}_1, \mathbf{m}_2, \ldots, \mathbf{m}_{N_m}\}$), where $\mathbf{S} = \mathbf{F} \cup \mathbf{M}$ ($\mathbf{S} = \{\mathbf{s}_1, \mathbf{s}_2, \ldots, \mathbf{s}_N\}$), such that $\mathbf{S} = \{\mathbf{s}_1 = \mathbf{f}_1, \mathbf{s}_2 = \mathbf{f}_2, \ldots, \mathbf{s}_{N_f} = \mathbf{f}_{N_f}, \mathbf{s}_{N_f+1} = \mathbf{m}_1, \mathbf{s}_{N_f+2} = \mathbf{m}_2, \ldots, \mathbf{s}_N = \mathbf{m}_{N_m}\}$.

2.3.1 Fitness Assignation

In the biological metaphor, the spider size is the characteristic that evaluates the individual capacity to perform better its assigned tasks. In the presented approach, every individual (spider) receive a weight w_i which represents the solution quality that corresponds to the spider i (irrespective of the gender) of the population S. In order to calculate the mass of every spider the next equations are used:

$$w_i = \frac{J(\mathbf{s}_i) - worst_{\mathbf{S}}}{best_{\mathbf{S}} - worst_{\mathbf{S}}} \tag{2.3}$$

where $J(\mathbf{s}_i)$ is the fitness value obtained by the evaluation of the spider position $\mathbf{s}_i$ with regard to the objective function $J(\cdot)$. The values $worst_{\mathbf{S}}$ and $best_{\mathbf{S}}$ are defined as follows (considering a maximization problem):

$$best_{\mathbf{S}} = \max_{k \in \{1,2,\ldots,N\}} (J(\mathbf{s}_k)) \text{ and } worst_{\mathbf{S}} = \min_{k \in \{1,2,\ldots,N\}} (J(\mathbf{s}_k)) \tag{2.4}$$

2.3.2 Modeling of the Vibrations Through the Communal Web

The communal web is used as a mechanism to transmit information among the colony members. This information, encoded in form of small vibrations, is a critical aspect for the collective coordination for all individuals in the population. The vibrations depend on the weight and distance of the spider which provoke them. Since the distance is relative to the individual that provokes the vibrations and the member who detects them, members near to the individual that provokes the vibrations perceive stronger vibrations in comparison with members located in distant positions. In order to reproduce this process, the vibrations perceived by the individual i as a result of the information transmitted by the member j are modeled according to the following equation:

$$Vib_{i,j} = w_j \cdot e^{-d_{i,j}^2} \tag{2.5}$$

where the $d_{i,j}$ is the Euclidian distance between the spiders i and j, such that $d_{i,j} = \left\| \mathbf{s}_i - \mathbf{s}_j \right\|$.

Although it is virtually possible to compute the perceived-vibrations considering any pair of individuals, three special relations are considered in the SSO approach:

1. The vibrations $Vibc_i$ perceived by the individual i ($\mathbf{s}_i$) as a result of the information transmitted by the member c ($\mathbf{s}_c$). Where c is an individual that has two important characteristics, it is the nearest member to i and posses a higher weight in comparison to $i(w_c > w_i)$.

$$Vibc_i = w_c \cdot e^{-d_{i,c}^2} \tag{2.6}$$

2. The vibrations $Vibb_i$ perceived by the individual i as a result of the information transmitted by the member b ($\mathbf{s}_b$). Where b is the individual with the best weight (best fitness value) of the entire population $\mathbf{S}$, such that $w_b = \max_{k \in \{1,2,\ldots,N\}} (w_k)$.

$$Vibb_i = w_b \cdot e^{-d_{i,b}^2} \tag{2.7}$$

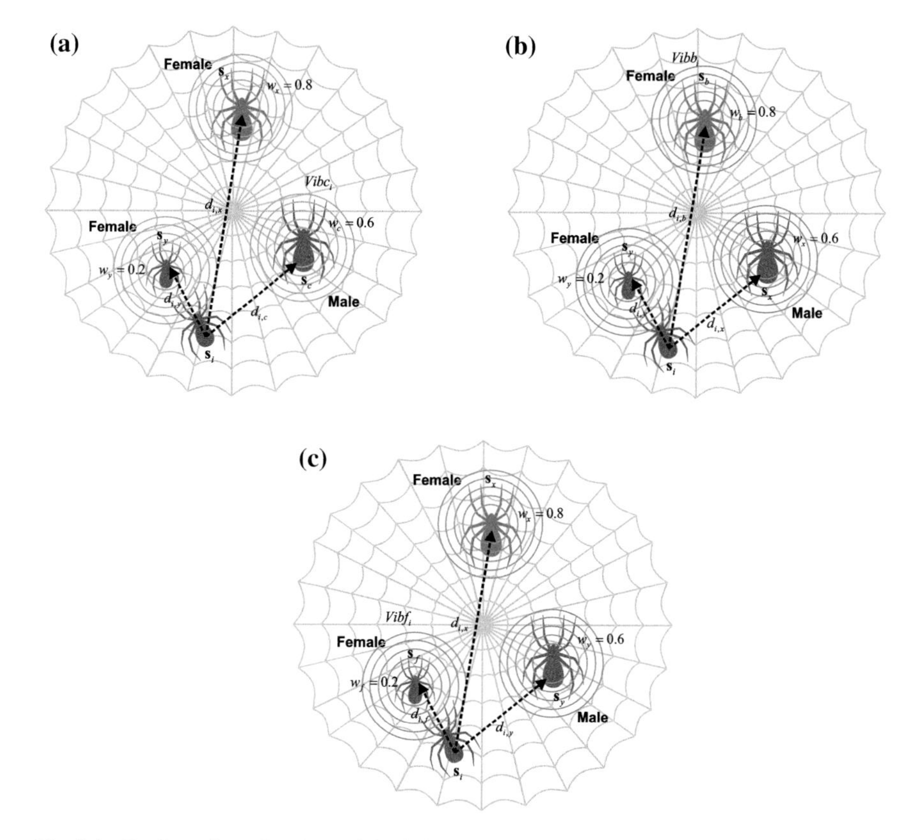

Fig. 2.1 Configuration of each special relation: **a** $Vibc_i$, **b** $Vibb_i$ and **c** $Vibf_i$

3. The vibrations $Vibf_i$ perceived by the individual i ($\mathbf{s}_i$) as a result of the information transmitted by the member f ($\mathbf{s}_f$). Where f is the nearest female individual to i.

$$Vibf_i = w_f \cdot e^{-d_{i,f}^2} \tag{2.8}$$

Figure 2.1 shows the configuration of each special relation: (a) $Vibc_i$, (b) $Vibb_i$ and (c) $Vibf_i$.

2.3.3 Initializing the Population

Like other evolutionary algorithms, the SSO is an iterative process; where the first step is to randomly initialize the entire population (females and males). The algorithm begins by initializing the set S of N spider positions. Each spider position, $\mathbf{f}_i$ or $\mathbf{m}_i$, is a n-dimensional vector containing the parameter values to be optimized. Such values are randomly and uniformly distributed between the pre-specified lower initial parameter bound p_j^{low} and the upper initial parameter bound p_j^{high}, just as it described by the following expressions:

$$f_{i,j}^0 = p_j^{low} + \mathrm{rand}(0,1) \cdot (p_j^{high} - p_j^{low}) \quad m_{k,j}^0 = p_j^{low} + \mathrm{rand}(0,1) \cdot (p_j^{high} - p_j^{low})$$
$$i = 1,2,\ldots,N_f; j = 1,2,\ldots,n \qquad k = 1,2,\ldots,N_m; j = 1,2,\ldots,n \tag{2.8}$$

where j, i and k are the parameter and individual indexes respectively whereas zero indicates the initial population. Hence, $f_{i,j}$ is the jth parameter of the ith female spider position.

2.3.4 *Cooperative Operators*

2.3.4.1 Female Cooperative Operator

Social-spiders perform cooperative interaction over other colony members. The way in which this behavior takes place depends on the spider gender. Female spiders present an attraction or dislike over other spiders irrespective of the gender. For a particular female spider, such attraction or dislike is commonly developed over other spiders that according to their vibrations (emitted over the communal web) represent strong colony members. Since the vibrations depend on the weight and distance of the members which provoke them, strong vibrations are produced either by big spiders or neighboring members relative to the individual which perceives them. The final decision of attraction or dislike over a determined member is taken according to an internal state which is influenced by several factors such as reproduction cycle, curiosity, and other random phenomena.

In order to emulate the cooperative behavior of the female spider, a new operator is defined. The operator considers the position change of the female spider i at each iteration. Such position change (which can be of attraction or repulsion) is computed as a combination of three different elements. The first one involves the change in regard to the nearest member to i with a higher weight (this member produces the vibration $Vibc_i$). The second one considers the change regarding the best individual of the entire population **S** (such individual produces the vibration $Vibb_i$). Finally, the third one implements the incorporation of a random movement. Since the final movement of attraction or repulsion depends on several random phenomena, this election is modeled as a stochastic decision. For this operation, a uniform random number r_m is generated within the range [0, 1]. If r_m is smaller than a threshold PF, an attraction movement is generated; otherwise, a repulsion movement is produced. Therefore, such operator can be modeled as follows:

$$\mathbf{f}_i^{k+1} = \begin{cases} \mathbf{f}_i^k + \alpha \cdot Vibc_i \cdot (\mathbf{s}_c - \mathbf{f}_i^k) + \beta \cdot Vibb_i \cdot (\mathbf{s}_b - \mathbf{f}_i^k) + \delta \cdot (\text{rand} - \frac{1}{2}) & \text{with probability } PF \\ \mathbf{f}_i^k - \alpha \cdot Vibc_i \cdot (\mathbf{s}_c - \mathbf{f}_i^k) - \beta \cdot Vibb_i \cdot (\mathbf{s}_b - \mathbf{f}_i^k) + \delta \cdot (\text{rand} - \frac{1}{2}) & \text{with probability } 1 - PF \end{cases} \quad (2.9)$$

where α, β, δ and rand are random numbers between [0, 1] whereas k represents the iteration number. The individual $\mathbf{s}_c$ and $\mathbf{s}_b$ represent the nearest member to i with a higher weight and the best individual of the entire population $\mathbf{S}$, respectively.

Under this operation, each particle presents a movement which combines the past position, with the attraction or repulsion vector over the local best element $\mathbf{s}_c$ and the global best individual seen $\mathbf{s}_b$ so-far. This particular type of interaction avoids the quick concentration of particles in only one point and encourages each particle to search around a local candidate region in its neighborhood ($\mathbf{s}_c$), rather than interacting with a particle ($\mathbf{s}_b$) in a distant region of the domain. The use of this scheme has two advantages. First, it prevents the particles from moving toward the global best position making the algorithm less susceptible to premature convergence. Second, it encourages the particles to explore their own neighborhood thoroughly before converging toward global best position. Therefore, it provides the algorithm with global search ability and enhances the exploitative behavior of the presented approach.

2.3.4.2 Male Cooperative Operator

According to the biological behavior of the social-spider, male population is divided in two classes: dominant and non-dominant male spiders. Dominant male spiders have better fitness characteristics (normally size) in comparison with non-dominant. Dominant males are attracted to the closest female spider in the communal web. In contrast, non-dominant male spiders tend to concentrate in the center of the male population, as a strategy to take advantage of the resources wasted by the dominant males.

For emulating such cooperative behavior, the male members are divided in two different groups (dominant members D and non-dominant members ND) according to their position with regard to the median member. Male members, with a weight value above the median value within the male population, are considered the dominant individuals D. On the other hand, the individuals under the median value are labeled as non-dominant ND males. In order to implement such computation, the male population $\mathbf{M}$ ($\mathbf{M} = \{\mathbf{m}_1, \mathbf{m}_2, \ldots, \mathbf{m}_{N_m}\}$) is arranged according to their weight value, in a decreasing order. Thus, the individual whose weight w_{N_f+m} is located in the middle is considered the median male member. Since the indexes of the male population $\mathbf{M}$ in regard to the entire population S are incremented by the number of female members N_f, the median weight is indexed by $N_f + m$. According to this, change of positions for the male spider can be modeled as follows:

$$\mathbf{m}_i^{k+1} = \begin{cases} \mathbf{m}_i^k + \alpha \cdot Vibf_i \cdot (\mathbf{s}_f - \mathbf{m}_i^k) + \delta \cdot (\text{rand} - \frac{1}{2}) & \text{if } w_{N_f+i} > w_{N_f+m} \\ \mathbf{m}_i^k + \alpha \cdot \left(\frac{\sum_{h=1}^{N_m} \mathbf{m}_h^k \cdot w_{N_f+h}}{\sum_{h=1}^{N_m} w_{N_f+h}} - \mathbf{m}_i^k \right) & \text{if } w_{N_f+i} \le w_{N_f+m} \end{cases} \tag{2.10}$$

where the individual $\mathbf{s}_f$ represents the nearest female individual to the male member i whereas $\left(\sum_{h=1}^{N_m} \mathbf{m}_h^k \cdot w_{N_f+h} / \sum_{h=1}^{N_m} w_{N_f+h}\right)$ correspond to the weighted mean of the male population $\mathbf{M}$.

By using this operator, two different behaviors are produced. First, the set $\mathbf{D}$ of particles is attracted to others in order to provoke mating. Such behavior allows to incorporate diversity in the population. Second, the set $\mathbf{ND}$ of particles are attracted to weighted mean of the male population $\mathbf{M}$. This fact is used to partially control the search process according to the average performance of a sub-group of the population. Such mechanism acts as a filter which avoids that very god individuals or extremely bad individuals influence the search process.

2.3.5 *Mating Operator*

Mating in a social-spider colony is performed by dominant males and the female members. Under such circumstances, when a dominant male $\mathbf{m}_g$ spider ($g \in \mathbf{D}$) locates a set $\mathbf{E}^g$ of female members within a specific range r (range of mating), it mates, forming a new brood $\mathbf{s}_{new}$ which is generated considering all the elements of the set $\mathbf{T}^g$ that, in turn, has been generated by the union $\mathbf{E}^g \cup \mathbf{m}_g$. It is important to emphasize that if the set $\mathbf{E}^g$ is empty, the mating operation is canceled. The range r is defined as a radius which depends on the size of the search space. Such radius r is computed according to the following model:

$$r = \frac{\sum_{j=1}^{n} (p_j^{high} - p_j^{low})}{2 \cdot n} \tag{2.11}$$

In the mating process, the weight of each involved spider (elements of $\mathbf{T}^g$) defines the probability of influence of each individual into the new brood. The spiders with heavier weight are more probable to influence the new product, while elements with lighter weight have a lower probability. The influence probability Ps_i of each member is assigned using the roulette method, which is defined as follows:

$$Ps_i = \frac{w_i}{\sum_{j \in \mathbf{T}^k} w_j}, \tag{2.12}$$

where $i \in \mathbf{T}^g$.

Once the new spider is formed, it is compared, the new spider candidate $\mathbf{s}_{new}$ with the worst spider $\mathbf{s}_{wo}$ of the colony, according to their weight values (where $w_{wo} = \min_{l \in \{1,2,\ldots,N\}} (w_l)$). If the new spider is better than the worst spider, the worst spider is replaced by the new one. Otherwise, the new spider is discarded and the population does not suffer changes. In case of replacement, the new spider assumes the sex and the same index of the replaced spider. Such fact assures that the entire population $\mathbf{S}$ maintains the original rate between female and male members.

In order to illustrate the mating operation, it is considered an example, where Fig. 2.2a is used as optimization problem. It is also assumed a population $\mathbf{S}$ of seven different 2-dimensional members ($N = 8$), five females ($N_f = 5$) and two males ($N_m = 3$). Figure 2.2b shows the initial configuration of the presented example. In the example, three different female members $\mathbf{f}_2(\mathbf{s}_2), \mathbf{f}_3(\mathbf{s}_3)$ and $\mathbf{f}_4(\mathbf{s}_4)$ constitute the set $\mathbf{E}^2$ located inside of the influence range r of a dominant male $\mathbf{m}_2(\mathbf{s}_7)$. Then, the new candidate spider $\mathbf{s}_{new}$ is generated from the elements $\mathbf{f}_2$, $\mathbf{f}_3$, $\mathbf{f}_4$ and $\mathbf{m}_2$ which constitute the set $\mathbf{T}^2$. Therefore, the value of the first decision variable $s_{new,1}$ for the new spider is chosen by means of the roulette mechanism considering the values already existing from the set $\{f_{2,1}, f_{3,1}, f_{4,1}, m_{2,1}\}$. The value of the second decision variable $s_{new,2}$ is also chosen in the same manner. Table 2.1 shows the data for

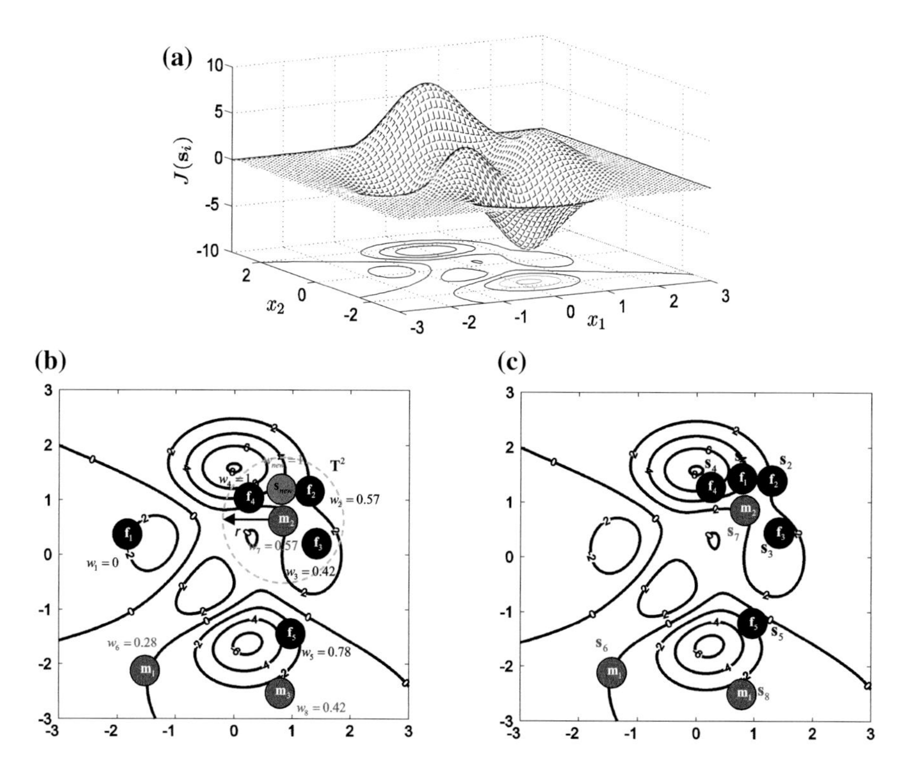

Fig. 2.2 Example of the mating operation: **a** optimization problem, **b** initial configuration before mating and **c** configuration after the mating operation

Table 2.1 Data for constructing the new spider $\mathbf{s}_{new}$ through the roulette method

Spider		Position	w_i	Ps_i	Roulette
$\mathbf{s}_1$	$\mathbf{f}_1$	(−1.9, 0.3)	0.00	–	m2 22%, f2 22%, f3 16%, f4 39%
$\mathbf{s}_2$	$\mathbf{f}_2$	(1.4, 1.1)	0.57	0.22	
$\mathbf{s}_3$	$\mathbf{f}_3$	(1.5, 0.2)	0.42	0.16	
$\mathbf{s}_4$	$\mathbf{f}_4$	(0.4, 1.0)	1.00	0.39	
$\mathbf{s}_5$	$\mathbf{f}_5$	(1.0, −1.5)	0.78	–	
$\mathbf{s}_6$	$\mathbf{m}_1$	(−1.3, −1.9)	0.28	–	
$\mathbf{s}_7$	$\mathbf{m}_2$	(0.9, 0.7)	0.57	0.22	
$\mathbf{s}_8$	$\mathbf{m}_3$	(0.8, −2.6)	0.42	–	
$\mathbf{s}_{new}$		(0.9, 1.1)	1.00	–	

constructing the new spider through the roulette method. Once the new spider $\mathbf{s}_{new}$ is formed, it is calculated its weight w_{new}. As $\mathbf{s}_{new}$ is better than the worst member $\mathbf{f}_1$ present in the population $\mathbf{S}$, $\mathbf{f}_1$ is replaced by $\mathbf{s}_{new}$. With the replacement, $\mathbf{s}_{new}$ assumes the same sex and index than $\mathbf{f}_1$. Figure 2.2c shows the configuration of $\mathbf{S}$ after the mating process.

Under this operation, new generated particles exploit locally the search space inside of the range of mating in order to find better individuals.

2.3.6 Computational Procedure

The computational procedure for the presented algorithm can be summarized as follows:

Step 1	Considering N as the total number of n-dimensional colony members, define the number of male N_m and females N_f in the entire population S $N_f = \text{floor}[(0.9 - \text{rand} \cdot 0.25) \cdot N]$ and $N_m = N - N_f$, where rand is a random number between [0, 1] whereas floor(·) maps a real number to an integer number
Step 2	Initialize randomly the female ($\mathbf{F} = \{\mathbf{f}_1, \mathbf{f}_2, \ldots, \mathbf{f}_{N_f}\}$) and male ($\mathbf{M} = \{\mathbf{m}_1, \mathbf{m}_2, \ldots, \mathbf{m}_{N_m}\}$) members where $\mathbf{S} = \{\mathbf{s}_1 = \mathbf{f}_1, \mathbf{s}_2 = \mathbf{f}_2, \ldots, \mathbf{s}_{N_f} = \mathbf{f}_{N_f}, \mathbf{s}_{N_f+1} = \mathbf{m}_1, \mathbf{s}_{N_f+2} = \mathbf{m}_2, \ldots, \mathbf{s}_N = \mathbf{m}_{N_m}\}$ and calculate the range of mating $r = \frac{\sum_{j=1}^{n}(p_j^{high} - p_j^{low})}{2 \cdot n}$ for $(i = 1; i < N_f + 1; i{+}{+})$ for $(j = 1; j < \mathrm{n} + 1; j{+}{+})$ $f_{i,j}^0 = p_j^{low} + \text{rand}(0,1) \cdot (p_j^{high} - p_j^{low})$ end for end for for $(k = 1; k < N_m + 1; k{+}{+})$ for $(j = 1; j < \mathrm{n} + 1; j{+}{+})$ $m_{k,j}^0 = p_j^{low} + \text{rand} \cdot (p_j^{high} - p_j^{low})$ end for end for
Step 3	Calculate the weight of every spider of $\mathbf{S}$ (Sect. 2.3.1) for $(i = 1, i < N + 1; i{+}{+})$ $w_i = \frac{J(\mathbf{s}_i) - worst_{\mathbf{S}}}{best_{\mathbf{S}} - worst_{\mathbf{S}}}$ where $best_{\mathbf{S}} = \max_{k \in \{1,2,\ldots,N\}}(J(\mathbf{s}_k))$ and $worst_{\mathbf{S}} = \min_{k \in \{1,2,\ldots,N\}}(J(\mathbf{s}_k))$ end for

(continued)

(continued)

Step 4	Move female spiders according to the female cooperative operator (Sect. 2.3.4)
	for $i = 1; i < N_f + 1; i{+}{+}$
	Calculate $Vibc_i$ and $Vibb_i$ (Sect. 2.3.2)
	If ($r_m < PF$); where $r_m \in \text{rand}(0,1)$
	$\mathbf{f}_i^{k+1} = \mathbf{f}_i^k + \alpha \cdot Vibc_i \cdot (\mathbf{s}_c - \mathbf{f}_i^k) + \beta \cdot Vibb_i \cdot (\mathbf{s}_b - \mathbf{f}_i^k) + \delta \cdot (\text{rand} - \frac{1}{2})$
	else if
	$\mathbf{f}_i^{k+1} = \mathbf{f}_i^k - \alpha \cdot Vibc_i \cdot (\mathbf{s}_c - \mathbf{f}_i^k) - \beta \cdot Vibb_i \cdot (\mathbf{s}_b - \mathbf{f}_i^k) + \delta \cdot (\text{rand} - \frac{1}{2})$
	end if
	end for

Step 5	Move the male spiders according to the male cooperative operator (Sect. 3.1.4)
	Find the median male individual (w_{N_f+m}) from **M**
	for ($i = 1; i < N_m + 1; i{+}{+}$)
	Calculate $Vibf_i$ (Sect. 2.3.2)
	If ($w_{N_f+i} > w_{N_f+m}$)
	$\mathbf{m}_i^{k+1} = \mathbf{m}_i^k + \alpha \cdot Vibf_i \cdot (\mathbf{s}_f - \mathbf{m}_i^k) + \delta \cdot (\text{rand} - \frac{1}{2})$
	else if
	$\mathbf{m}_i^{k+1} = \mathbf{m}_i^k + \alpha \cdot \left(\frac{\sum_{h=1}^{N_m} \mathbf{m}_h^k \cdot w_{N_f+h}}{\sum_{h=1}^{N_m} w_{N_f+h}} - \mathbf{m}_i^k \right)$
	end if
	end for

Step 6	Perform mating operation (Sect. 2.3.5)
	for $i = 1; i < N_m + 1; i{+}{+}$
	If ($\mathbf{m}_i \in \mathbf{D}$)
	Find $\mathbf{E}^i$
	If ($\mathbf{E}^i$ is not empty)
	Form $\mathbf{s}_{new}$ using the roulette method
	If ($w_{new} > w_{wo}$)
	$\mathbf{s}_{wo} = \mathbf{s}_{new}$
	end if
	end if
	end if
	end for

Step 7	If the stop criteria is met, the process is finished; otherwise, go back to Step 3

2.3.7 Discussion About the SSO Algorithm

Evolutionary algorithms (EA) have been widely employed for solving complex optimization problems. These methods are found to be more powerful than conventional methods based on formal logics or mathematical programming [32]. In an EA algorithm, search agents have to decide whether to explore unknown search positions or to exploit already tested positions in order to improve their solution quality. Pure exploration degrades the precision of the evolutionary process but increases its capacity to find new potentially solutions. On the other hand, pure exploitation allows refining existent solutions but adversely drives the process to local optimal solutions. Therefore, the ability of an EA to find a global optimal solution depends on its capacity to find a good balance between the exploitation of found-so-far elements and the exploration of the search space [33]. So far, the exploration–exploitation dilemma has been an unsolved issue within the framework of evolutionary algorithms.

EA define individuals with the same property, performing virtually the same behavior. Under these circumstances, algorithms waste the possibility to add new and selective operators as a result of considering individuals with different characteristics. These operators could incorporate computational mechanisms to improve several important algorithm characteristics such as population diversity or searching capacities.

On the other hand, PSO and ABC are the most popular swarm algorithms for solving complex optimization problems. However, they present serious flaws such as premature convergence and difficulty to overcome local minima [10, 11]. The reason of these problems is the operators used for modifying the individual positions. In such algorithms, during their evolution, the position of each agent in the next iteration is updated yielding an attraction towards the position of the best particle seen so-far (in case of PSO) or of other randomly chosen individual (in case of ABC). Such behaviors produce that the entire population, as the algorithm evolves, concentrates around the best particle or diverges without control, favoring the premature convergence or damaging the exploration-exploitation balance [12, 13].

Different to other EA, in SSO, each individual is modeled considered two different genders. Such fact allows incorporating computational mechanisms to avoid critical flaws present in the popular PSO and ABC algorithms, such as premature convergence and incorrect exploration-exploitation balance. Since the optimization point of view, the use of the social-spider behavior as metaphor introduces interesting concepts in the evolutionary algorithms: The fact of dividing the entire population in different search-agent categories and the employment of specialized operators applied selectively to each of them. Using this framework, it is possible to improve the balance between exploitation and exploration, conserving in the same population, individuals who achieve efficient exploration (female spiders) and individuals that verify extensive exploitation (male spiders). Furthermore, the social-spider behavior mechanism introduces an interesting computational scheme. Such scheme presents three important particularities. First, individuals are separately processed according to their characteristics. Second, the operators share the same communication mechanism. This mechanism allows employing important information of the evolutionary process to modify the influence of each operator. Third, although the operators modify the position of only an individual type, they use global information (positions of all individual types) in order to perform the modification. Figure 2.3 presents a schematic representation of the algorithm-data-flow. According to Fig. 2.3, the female cooperative and male cooperative operators process only female or male individuals, respectively. However, the mating operator modifies both individual types.

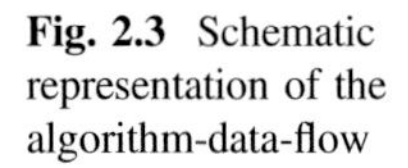
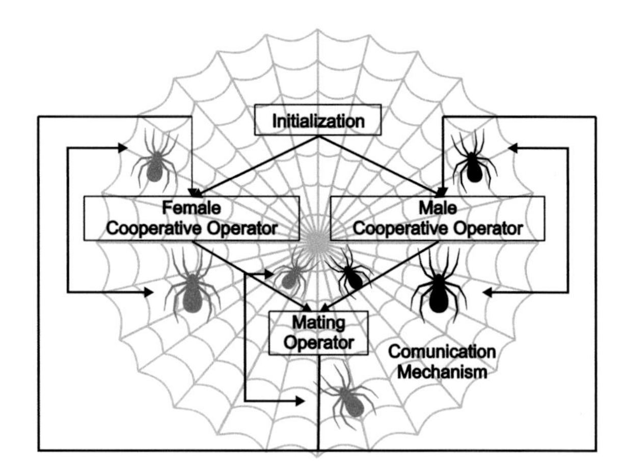

Fig. 2.3 Schematic representation of the algorithm-data-flow

2.4 Experimental Results

A comprehensive set of 19 functions, collected from Refs. [34–40], has been used to test the performance of the presented approach. Table 2.2 present the benchmark functions used in our experimental study. In the table, n indicates the dimension of the function, $f(\mathbf{x}^*)$ the optimum value of the function, $\mathbf{x}^*$ the optimum position and S the search space (subset of R^n). A detailed description of each function is given in Table 2.2.

2.4.1 Performance Comparison to Other Swarm Algorithms

The SSO algorithm was applied to 19 functions whose results have been compared to those produced by the Particle Swarm Optimization (PSO) method [3] and the Artificial Bee Colony (ABC) algorithm [4]. These are considered as the most popular swarm algorithms in many optimization applications. In all comparisons, the population has been set to 50 individuals. The maximum iteration number for all functions has been set to 1000. Such stop criterion has been selected to maintain compatibility to similar works reported in the literature [41, 42].

The parameter setting for each algorithm in the comparison is described as follows:

1. PSO [3]: The parameters are set to $c_1 = 2$ and $c_2 = 2$; besides, the weight factor decreases linearly from 0.9 to 0.2.
2. ABC: The algorithm has been implemented using the guidelines provided by its own reference [4], using the parameter $limit = 100$.
3. SSO: Once determined experimentally, the parameter PF was set to 0.7. It is kept for all experiments presented in this section.

Table 2.2 Test functions used in the experimental study

Name	Function	S	Dim	Minimum
Sphere	$f_1(\mathbf{x}) = \sum_{i=1}^{n} x_i^2$	$[-100, 100]^n$	$n = 30$	$\mathbf{x}^* = (0, \ldots, 0)$; $f(\mathbf{x}^*) = 0$
Schwefel 2.22	$f_2(\mathbf{x}) = \sum_{i=1}^{n} \lvert x_i \rvert + \prod_{i=1}^{n} \lvert x_i \rvert$	$[-10, 10]^n$	$n = 30$	$\mathbf{x}^* = (0, \ldots, 0)$; $f(\mathbf{x}^*) = 0$
Schwefel 1.2	$f_3(\mathbf{x}) = \sum\limits_{i=1}^{n} \left(\sum_{j=1}^{i} x_j \right)^2$	$[-100, 100]^n$	$n = 30$	$\mathbf{x}^* = (0, \ldots, 0)$; $f(\mathbf{x}^*) = 0$
F4	$f_4(\mathbf{x}) = 418.9829n + \sum_{i=1}^{n} \left(-x_i \sin\left(\sqrt{\lvert x_i \rvert} \right) \right)$	$[-100, 100]^n$	$n = 30$	$\mathbf{x}^* = (0, \ldots, 0)$; $f(\mathbf{x}^*) = 0$
Rosenbrock	$f_5(\mathbf{x}) = \sum_{i=1}^{n-1} \left[100(x_{i+1} - x_i^2)^2 + (x_i - 1)^2 \right]$	$[-30, 30]^n$	$n = 30$	$\mathbf{x}^* = (1, \ldots, 1)$; $f(\mathbf{x}^*) = 0$
Step	$f_6(\mathbf{x}) = \sum_{i=1}^{n} \left(\lfloor x_i + 0.5 \rfloor \right)^2$	$[-100, 100]^n$	$n = 30$	$\mathbf{x}^* = (0, \ldots, 0)$; $f(\mathbf{x}^*) = 0$
Quartic	$f_7(\mathbf{x}) = \sum_{i=1}^{n} i x_i^4 + random(0, 1)$	$[-1.28, 1.28]^n$	$n = 30$	$\mathbf{x}^* = (0, \ldots, 0)$; $f(\mathbf{x}^*) = 0$
Dixon and price	$f_8(\mathbf{x}) = (x_1 - 1)^2 + \sum_{i=1}^{n} i \left(2x_i^2 - x_{i-1} \right)^2$	$[-10, 10]^n$	$n = 30$	$\mathbf{x}^* = (0, \ldots, 0)$; $f(\mathbf{x}^*) = 0$

(continued)

Table 2.2 (continued)

Name	Function	S	Dim	Minimum
Levy	$f_9(\mathbf{x}) = 0.1\begin{cases} \sin^2(3\pi x_1) \\ + \sum_{i=1}^{n}(x_i-1)^2\left[1+\sin^2(3\pi x_i+1)\right] \\ +(x_n-1)^2\left[1+\sin^2(2\pi x_n)\right] \end{cases}$ $+ \sum_{i=1}^{n} u(x_i, 5, 100, 4);$ $u(x_i,a,k,m) = \begin{cases} k(x_i-a)^m & x_i > a \\ 0 & -a < x_i < a \\ k(-x_i-a)^m & x_i < -a \end{cases}$	$[-10, 10]^n$	$n = 30$	$\mathbf{x}^* = (1,\ldots,1)$; $f(\mathbf{x}^*) = 0$
Sum of Squares	$f_{10}(\mathbf{x}) = \sum_{i=1}^{n} i x_i^2$	$[-10, 10]^n$	$n = 30$	$\mathbf{x}^* = (0,\ldots,0)$; $f(\mathbf{x}^*) = 0$
Zakharov	$f_{11}(\mathbf{x}) = \sum_{i=1}^{n} x_i^2 + \left(\sum_{i=1}^{n} 0.5 i x_i\right)^2 + \left(\sum_{i=1}^{n} 0.5 i x_i\right)^4$	$[-5, 10]^n$	$n = 30$	$\mathbf{x}^* = (0,\ldots,0)$; $f(\mathbf{x}^*) = 0$
Penalized	$f_{12}(\mathbf{x}) = \frac{\pi}{n}\begin{Bmatrix} 10\sin(\pi y_1) + \\ \sum_{i=1}^{n-1}(y_i-1)^2\left[1+10\sin^2(\pi y_{i+1})\right] + (y_n-1)^2 \end{Bmatrix}$ $+ \sum_{i=1}^{n} u(x_i, 10, 100, 4)$ $y_i = 1 + \frac{(x_i+1)}{4}$ $u(x_i,a,k,m) = \begin{cases} k(x_i-a)^m & x_i > a \\ 0 & -a \le x_i \le a \\ k(-x_i-a)^m & x_i < a \end{cases}$	$[-50, 50]^n$	$n = 30$	$\mathbf{x}^* = (0,\ldots,0)$; $f(\mathbf{x}^*) = 0$

(continued)

Table 2.2 (continued)

Name	Function	S	Dim	Minimum
Penalized 2	$f_{13}(\mathbf{x}) = 0.1\left\{ \begin{array}{l} \sin^2(3\pi x_1) \\ + \sum_{i=1}^{n} \begin{array}{l} (x_i - 1)^2\left[1 + \sin^2(3\pi x_i + 1)\right] \\ + (x_n - 1)^2\left[1 + \sin^2(2\pi x_n)\right] \end{array} \end{array} \right\}$ $+ \sum_{i=1}^{n} u(x_i, 5, 100, 4)$ where $u(x_i, a, k, m)$ is the same as Penalized function	$[-50, 50]^n$	$n = 30$	$\mathbf{x}^* = (0, \ldots, 0)$; $f(\mathbf{x}^*) = 0$
Schwefel	$f_{14}(\mathbf{x}) = \sum_{i=1}^{n} -x_i \sin\left(\sqrt{\lvert x_i \rvert}\right)$	$[-500, 500]^n$	$n = 30$	$\mathbf{x}^* = (420, \ldots, 420)$; $f(\mathbf{x}^*) = -418.9829 \times n$
Rastrigin	$f_{15}(\mathbf{x}) = \sum_{i=1}^{n} \left[x_i^2 - 10\cos(2\pi x_i) + 10\right]$	$[-5.12, 5.12]^n$	$n = 30$	$\mathbf{x}^* = (0, \ldots, 0)$; $f(\mathbf{x}^*) = 0$
Ackley	$f_{16}(\mathbf{x}) = -20\exp\left(-0.2\sqrt{\frac{1}{n}\sum_{i=1}^{n} x_i^2}\right)$ $- \exp\left(\frac{1}{n}\sum_{i=1}^{n}\cos(2\pi x_i)\right) + 20 + \exp$	$[-32, 32]^n$	$n = 30$	$\mathbf{x}^* = (0, \ldots, 0)$; $f(\mathbf{x}^*) = 0$
Griewank	$f_{17}(\mathbf{x}) = \frac{1}{4000}\sum_{i=1}^{n} x_i^2 - \prod_{i=1}^{n}\cos\left(\frac{x_i}{\sqrt{i}}\right) + 1$	$[-600, 600]^n$	$n = 30$	$\mathbf{x}^* = (0, \ldots, 0)$; $f(\mathbf{x}^*) = 0$
Powelll	$f_{18}(\mathbf{x}) = \sum_{i=1}^{n/k}(x_{4i-3} + 10x_{4i-2})^2 + 5(x_{4i-1} - x_{4i})^2$ $+ (x_{4i-2} - x_{4i-1})^4 + 10(x_{4i-3} - x_{4i})^4$	$[-4, 5]^n$	$n = 30$	$\mathbf{x}^* = (0, \ldots, 0)$; $f(\mathbf{x}^*) = 0$
Salomon	$f_{19}(\mathbf{x}) = -\cos\left(2\pi\sqrt{\sum_{i=1}^{n} x_i^2}\right) + 0.1\sqrt{\sum_{i=1}^{n} x_i^2 + 1}$	$[-100, 100]^n$	$n = 30$	$\mathbf{x}^* = (0, \ldots, 0)$; $f(\mathbf{x}^*) = 0$

The experiment compares the SSO to other algorithms such as PSO and ABC. The results for 30 runs are reported in Table 2.3 considering the following performance indexes: the Average Best-so-far (AB) solution, the Median Best-so-far (MB) and the Standard Deviation (SD) of best-so-far solution. The best outcome for each function is boldfaced. According to this table, SSO delivers better results than PSO and ABC for all functions. In particular, the test remarks the largest difference in performance which is directly related to a better trade-off between exploration and exploitation.

A non-parametric statistical significance proof known as the Wilcoxon's rank sum test for independent samples [43, 44] has been conducted over the "average best-so-far" (AB) data of Table 2.2, with a 5 % significance level. Table 2.4 reports the p-values produced by Wilcoxon's test for the pair-wise comparison of the "average best so-far" of two groups. Such groups are formed by SSO vs. PSO and SSO vs. ABC. As a null hypothesis, it is assumed that there is no significant difference between mean values of the two algorithms. The alternative hypothesis

Table 2.3 Minimization result of benchmark functions of Table 2.2 with $n = 30$

		SSO	ABC	PSO
$f_1(x)$	AB	1.96E−03	2.90E−03	1.00E+03
	MB	2.81E−03	1.50E−03	2.08E−09
	SD	9.96E−04	1.44E−03	3.05E+03
$f_2(x)$	AB	1.37E−02	1.35E−01	5.17E+01
	MB	1.34E−02	1.05E−01	5.00E+01
	SD	3.11E−03	8.01E−02	2.02E+01
$f_3(x)$	AB	4.27E−02	1.13E+00	8.63E+04
	MB	3.49E−02	6.11E−01	8.00E+04
	SD	3.11E−02	1.57E+00	5.56E+04
$f_4(x)$	AB	5.40E−02	5.82E+01	1.47E+01
	MB	5.43E−02	5.92E+01	1.51E+01
	SD	1.01E−02	7.02E+00	3.13E+00
$f_5(x)$	AB	1.14E+02	1.38E+02	3.34E+04
	MB	5.86E+01	1.32E+02	4.03E+02
	SD	3.90E+01	1.55E+02	4.38E+04
$f_6(x)$	AB	2.68E−03	4.06E−03	1.00E+03
	MB	2.68E−03	3.74E−03	1.66E−09
	SD	6.05E−04	2.98E−03	3.06E+03
$f_7(x)$	AB	1.20E+01	1.21E+01	1.50E+01
	MB	1.20E+01	1.23E+01	1.37E+01
	SD	5.76E−01	9.00E−01	4.75E+00
$f_8(x)$	AB	2.14E+00	3.60E+00	3.12E+04
	MB	3.64E+00	8.04E−01	2.08E+02
	SD	1.26E+00	3.54E+00	5.74E+04

(continued)

Table 2.3 (continued)

		SSO	ABC	PSO
$f_9(x)$	AB	6.92E−05	1.44E−04	2.47E+00
	MB	6.80E−05	8.09E−05	9.09E−01
	SD	4.02E−05	1.69E−04	3.27E+00
$f_{10}(x)$	AB	4.44E−04	1.10E−01	6.93E+02
	MB	4.05E−04	4.97E−02	5.50E+02
	SD	2.90E−04	1.98E−01	6.48E+02
$f_{11}(x)$	AB	6.81E+01	3.12E+02	4.11E+02
	MB	6.12E+01	3.13E+02	4.31E+02
	SD	3.00E+01	4.31E+01	1.56E+02
$f_{12}(x)$	AB	5.39E−05	1.18E−04	4.27E+07
	MB	5.40E−05	1.05E−04	1.04E−01
	SD	1.84E−05	8.88E−05	9.70E+07
$f_{13}(x)$	AB	1.76E−03	1.87E−03	5.74E−01
	MB	1.12E−03	1.69E−03	1.08E−05
	SD	6.75E−04	1.47E−03	2.36E+00
$f_{14}(x)$	AB	−9.36E+02	−9.69E+02	−9.63E+02
	MB	−9.36E+02	−9.60E+02	−9.92E+02
	SD	1.61E+01	6.55E+01	6.66E+01
$f_{15}(x)$	AB	8.59E+00	2.64E+01	1.35E+02
	MB	8.78E+00	2.24E+01	1.36E+02
	SD	1.11E+00	1.06E+01	3.73E+01
$f_{16}(x)$	AB	1.36E−02	6.53E−01	1.14E+01
	MB	1.39E−02	6.39E−01	1.43E+01
	SD	2.36E−03	3.09E−01	8.86E+00
$f_{17}(x)$	AB	3.29E−03	5.22E−02	1.20E+01
	MB	3.21E−03	4.60E−02	1.35E−02
	SD	5.49E−04	3.42E−02	3.12E+01
$f_{18}(x)$	AB	1.87E+00	2.13E+00	1.26E+03
	MB	1.61E+00	2.14E+00	5.67E+02
	SD	1.20E+00	1.22E+00	1.12E+03
$f_{19}(x)$	AB	2.74E−01	4.14E+00	1.53E+00
	MB	3.00E−01	4.10E+00	5.50E−01
	SD	5.17E−02	4.69E−01	2.94E+00

Maximum number of iterations = 1000

considers a significant difference between the "average best-so-far" values of both approaches. All p-values reported in Table 2.4 are less than 0.05 (5 % significance level) which is a strong evidence against the null hypothesis. Therefore, such evidence indicates that SMS results are statistically significant and that it has not occurred by coincidence (i.e. due to common noise contained in the process).

Table 2.4 p-values produced by Wilcoxon's test comparing SSO versus ABC and SSO versus PSO over the "average best-so-far" (AB) values from Table 2.3

Function	SSO versus ABC	SSO versus PSO
$f_1(x)$	0.041	1.8E−05
$f_2(x)$	0.048	0.059
$f_3(x)$	5.4E−04	6.2E−07
$f_4(x)$	1.4E−07	4.7E−05
$f_5(x)$	0.045	7.1E−07
$f_6(x)$	2.3E−04	5.5E−08
$f_7(x)$	0.048	0.011
$f_8(x)$	0.017	0.043
$f_9(x)$	8.1E−04	2.5E−08
$f_{10}(x)$	4.6E−06	1.7E−09
$f_{11}(x)$	9.2E−05	7.8E−06
$f_{12}(x)$	0.022	1.1E−10
$f_{13}(x)$	0.048	2.6E−05
$f_{14}(x)$	0.044	0.049
$f_{15}(x)$	4.5E−05	7.9E−08
$f_{16}(x)$	2.8E−05	4.1E−06
$f_{17}(x)$	7.1E−04	6.2E−10
$f_{18}(x)$	0.013	8.3E−10
$f_{19}(x)$	4.9E−05	5.1E−08

2.5 Summary

In this chapter, a novel swarm algorithm, namely the Social Spider Optimization (SSO) has been presented for solving optimization tasks. The SSO algorithm is based on the simulation of the cooperative behavior of social-spiders. In the presented algorithm, individuals emulate a group of spiders which interact to each other based on the biological laws of the cooperative colony. The algorithm considers two different search agents (spiders): males and females. Depending on the sex, each individual is conducted by a set of different evolutionary operators which mimic the different cooperative behaviors assumed in the colony.

Different to most of the existent swarm algorithms, in the presented approach, each individual is modeled considered two different genders. Such fact allows not only to emulate in a better realistic way the cooperative behavior of the colony, but also to incorporate computational mechanisms to avoid critical flaws present in the popular PSO and ABC algorithms, such as premature convergence and incorrect exploration-exploitation balance.

SSO has been experimentally tested considering a suite of 19 benchmark functions. The performance of SSO has been also compared to the following swarm algorithms: the Particle Swarm Optimization method (PSO) [16], and the Artificial Bee Colony (ABC) algorithm [38]. Results have confirmed a high performance of the presented method in terms of the solution quality for solving most of benchmark functions.

The SSO's remarkable performance is associated with two different reasons: (i) the defined operators allow a better particle distribution in the search space, increasing the algorithm's ability to find the global optima; and (ii) the division of the population in different individual types, provides the use of different rates between exploration and exploitation during the evolution process.

References

1. Bonabeau, E., Dorigo, M., Theraulaz, G.: Swarm Intelligence: from Natural to Artificial Systems. Oxford University Press Inc, New York, NY, USA (1999)
2. Kassabalidis, I., El-Sharkawi, M.A., II Marks R.J., Arabshahi, P., Gray, A.A.: Swarm intelligence for routing in communication networks, Global Telecommunications Conference, GLOBECOM'01, vol. 6, pp. 3613–3617. IEEE (2001)
3. Kennedy, J., Eberhart, R.: Particle swarm optimization. In Proceedings of the 1995 IEEE International Conference on Neural Networks, vol. 4, pp. 1942–1948. (1995)
4. Karaboga, D.: An Idea Based on Honey Bee Swarm for Numerical Optimization. TechnicalReport-TR06. Erciyes University, Turkey (2005)
5. Passino, K.M.: Biomimicry of bacterial foraging for distributed optimization and control. IEEE Control Syst. Mag. **22**(3), 52–67 (2002)
6. Hossein, A., Hossein-Alavi, A.: Krill herd: a new bio-inspired optimization algorithm. Commun. Nonlinear Sci. Numer. Simul. **17**, 4831–4845 (2012)
7. Yang, X.S: Engineering optimization: An introduction with metaheuristic applications. Wiley (2010)
8. Rajabioun, R.: Cuckoo optimization algorithm. Appl. Soft Comput. **11**, 5508–5518 (2011)
9. Bonabeau, E.: Social insect colonies as complex adaptive systems. Ecosystems **1**, 437–443 (1998)
10. Wang, Y., Li, B., Weise, T., Wang, J., Yuan, B., Tian, Q.: Self-adaptive learning based particle swarm optimization. Inf. Sci. **181**(20), 4515–4538 (2011)
11. Wan-li, X., Mei-qing, A.: An efficient and robust artificial bee colony algorithm for numerical optimization. Comput. Oper. Res. **40**, 1256–1265 (2013)
12. Wang, H., Sun, H., Li, C., Rahnamayan, S., Jeng-shyang, P.: Diversity enhanced particle swarm optimization with neighborhood. Inf. Sci. **223**, 119–135 (2013)
13. Banharnsakun, A., Achalakul, T., Sirinaovakul, B.: The best-so-far selection in artificial bee colony algorithm. Appl. Soft Comput. **11**, 2888–2901 (2011)
14. Gordon, D.: The organization of work in social insect colonies. Complexity **8**(1), 43–46 (2003)
15. Lubin, T.B.: The evolution of sociality in spiders. In: Brockmann, H.J. (ed.) Advances in the study of behavior, vol. 37, pp. 83–145 (2007)
16. Uetz, G.W.: Colonial web-building spiders: Balancing the costs and. In: Choe, E.J., Crespi, B. (ed.) The Evolution of Social Behavior in Insects and Arachnids, pp. 458–475. Cambridge, Cambridge University Press, England
17. Aviles, L.: Sex-ratio bias and possible group selection in the social spider Anelosimus eximius. Am. Nat. **128**(1), 1–12 (1986)
18. Burgess, J. W.: Social spacing strategies in spiders. In: Rovner, P.N. (ed.) Spider Communication: Mechanisms and Ecological Significance, pp. 317–351. Princeton University Press, Princeton, New Jersey (1982)
19. Maxence, S.: Social organization of the colonial spider Leucauge sp. in the Neotropics: Vertical stratification within colonies. The. J. Arachnology **38**, 446–451 (2010)

20. Eric, C., Yip, K.S.: Cooperative capture of large prey solves scaling challenge faced by spider societies. Proc. Natl. Acad. Sci. USA **105**(33), 11818–11822 (2008)
21. Oster, G., Wilson, E.: Caste and ecology in the social insects. Princeton University Press, Princeton, N.J. (1978)
22. Hölldobler, Bert., Wilson, E.O.: Journey to the ants: a story of scientific exploration. ISBN 0-674-48525-4 (1994)
23. Hölldobler, Bert., Wilson, E.O.: The Ants. Harvard University Press,USA. ISBN 0-674-04075-9 (1990)
24. Avilés, L.: Causes and consequences of cooperation and permanent-sociality in spiders. In: Choe, B.C. (ed.) The Evolution of Social Behavior in Insects and Arachnids, pp. 476–498. Cambridge University Press, Cambridge, Massachusetts (1997)
25. Rayor, E.C.: Do social spiders cooperate in predator defense and foraging without a web? Behav. Ecol. Sociobiol. **65**(10), 1935–1945 (2011)
26. Gove, R., Hayworth, M., Chhetri, M., Rueppell, O.: Division of labour and social insect colony performance in relation to task and mating number under two alternative response threshold models. Insect. Soc. **56**(3), 19–331 (2009)
27. Ann, L., Rypstra, R.S.: Prey Size, prey perishability and group foraging in a social spider. Oecologia **86**(1), 25–30 (1991)
28. Pasquet, A.: Cooperation and prey capture efficiency in a social spider, Anelosimus eximius (Araneae, Theridiidae). Ethology **90**, 121–133 (1991)
29. Ulbrich, K., Henschel, J.: Intraspecific competition in a social spider. Ecol. Model. **115**(2–3), 243–251 (1999)
30. Jones, T., Riechert, S.: Patterns of reproductive success associated with social structure and microclimate in a spider system. Anim. Behav. **76**(6), 2011–2019 (2008)
31. Damian, O., Andrade, M., Kasumovic, M.: Dynamic population structure and the evolution of spider mating systems. Adv. Insect Physiol. **41**, 65–114 (2011)
32. Yang, X.-S.: Nature-inspired metaheuristic algorithms. Luniver Press, Beckington (2008)
33. Chen, D.B., Zhao, C.X.: Particle swarm optimization with adaptive population size and its application. Appl. Soft Comput. **9**(1), 39–48 (2009)
34. Storn, R., Price, K.: Differential evolution—a simple and efficient heuristicfor global optimization over continuous spaces. J. Global Optim. **11**(4), 341–359 (1995)
35. Yang, E, Barton, NH., Arslan, T., Erdogan, AT.: A novel shifting balance theory-based approach to optimization of an energy-constrained modulation scheme for wireless sensor networks. In: Proceedings of the IEEE Congress on Evolutionary Computation, CEC 2008, June 1–6, 2008, Hong Kong, China. pp. 2749–2756, IEEE (2008)
36. Duan, X., Wang, G.G., Kang, X., Niu, Q., Naterer, G., Peng, Q.: Performance study of mode-pursuing sampling method. Eng. Optim. 41(1) (2009)
37. Vesterstrom, J., Thomsen, R.: A comparative study of differential evolution, particle swarm optimization, and evolutionary algorithms on numerical benchmark problems. Evolutionary Computation, 2004. CEC2004. Congress on, vol. 2, pp. 1980–1987, 19–23 June 2004
38. Mezura-Montes, E., Velázquez-Reyes, J., Coello Coello, C.A.: A comparative study of differential evolution variants for global optimization, In: Proceedings of the 8th annual conference on Genetic and evolutionary computation (GECCO'06), pp. 485–492 ACM, New York, NY, USA (2006)
39. Karaboga, D., Akay, B.: A comparative study of Artificial Bee Colony algorithm, Applied Mathematics and Computation, vol. 214, Issue 1, 1 Aug 2009, pp. 108–132. ISSN 0096-3003 (2009)
40. Krishnanand, K.R., Nayak, S.K., Panigrahi, B.K., Rout, P.K.: Comparative study of five bio-inspired evolutionary optimization techniques, Nature & Biologically Inspired Computing, NaBIC, World Congress on, vol., pp. 1231–1236 (2009)
41. Ying, J., Ke-Cun, Z., Shao-Jian, Q.: A deterministic global optimization algorithm. Appl. Math. Comput. **185**(1), 382–387 (2007)

42. Rashedia, E., Nezamabadi-pour, H., Saryazdi, S.: Filter modeling using gravitational search algorithm. Eng. Appl. Artif. Intell. **24**(1), 117–122 (2011)
43. Wilcoxon, F.: Individual comparisons by ranking methods. Biometrics **1**, 80–83 (1945)
44. Garcia, S., Molina, D., Lozano, M., Herrera, F.: A study on the use of non-parametric tests for analyzing the evolutionary algorithms' behaviour: a case study on the CEC'2005 Special session on real parameter optimization. J Heurist (2008). doi:10.1007/s10732-008-9080-4

Chapter 3
A States of Matter Algorithm for Global Optimization

3.1 Introduction

Global optimization is a field with applications in many areas of science, engineering, economics and others, where mathematical modelling is used [1]. In general, the goal is to find a global optimum of an objective function defined over a given search space. Global optimization algorithms are usually broadly divided into deterministic and evolutionary [2]. Since deterministic methods only provide a theoretical guarantee of locating a local minimum of the objective function, they often face great difficulties in solving global optimization problems [3]. On the other hand, evolutionary algorithms are usually faster in locating a global optimum than deterministic ones [4]. Moreover, evolutionary methods adapt easily to black-box formulations and extremely ill-behaved functions, whereas deterministic methods usually rest on at least some theoretical assumptions about the problem formulation and its analytical properties (such as Lipschitz continuity) [5].

Several evolutionary algorithms have been developed by a combination of rules and randomness mimicking several natural phenomena. Such phenomena include evolutionary processes such as the Evolutionary Algorithm (EA) proposed by Fogel et al. [6], De Jong [7], and Koza [8], the Genetic Algorithm (GA) proposed by Holland [9] and Goldberg [10], the Artificial Immune System proposed by De Castro et al. [11] and the Differential Evolution Algorithm (DE) proposed by Price and Storn [12]. Some other methods which are based on physical processes include the Simulated Annealing proposed by Kirkpatrick et al. [13], the Electromagnetism-like Algorithm proposed by İlker et al. [14], the Gravitational Search Algorithm proposed by Rashedi et al. [15]. Also, there are other methods based on the animal-behavior phenomena such as the Particle Swarm Optimization (PSO) algorithm proposed by Kennedy and Eberhart [16] and the Ant Colony Optimization (ACO) algorithm proposed by Dorigo et al. [17].

Every EA needs to address the exploration and exploitation of a search space. Exploration is the process of visiting entirely new points of a search space, whilst

E. Cuevas et al., *Advances of Evolutionary Computation: Methods and Operators*, Studies in Computational Intelligence 629,
DOI 10.1007/978-3-319-28503-0_3

exploitation is the process of refining those points of a search space within the neighborhood of previously visited locations in order to improve their solution quality. Pure exploration degrades the precision of the evolutionary process but increases its capacity to find new potentially solutions. On the other hand, pure exploitation allows refining existent solutions but adversely drives the process to local optimal solutions. Therefore, the ability of an EA to find a global optimal solution depends on its capacity to find a good balance between the exploitation of found-so-far elements and the exploration of the search space [18]. So far, the exploration–exploitation dilemma has been an unsolved issue within the framework of evolutionary algorithms.

Although PSO, DE and GSA are considered the most popular algorithms in many optimization applications, they fail in finding a balance between exploration and exploitation [19]; in multimodal functions, they cannot explore the whole region effectively and often suffers from the problem of premature convergence or loss of diversity. In order to deal with this problem, several proposals have been suggested in the literature [20–38]. In most of the approaches, exploration and exploitation is modified by the proper settings of control parameters that have an influence on algorithm search capabilities [39]. One common strategy is that EAs should start with exploration and then gradually change into exploitation [40]. Such a policy can be easily described with deterministic approaches where the operator that controls the individual diversity decreases along with the evolution. This is generally correct, but such a policy tends to face difficulties when solving certain problems e.g., multimodal functions with many optima, since premature takeover of exploitation over exploration occurs. Some approaches that uses this strategy can be found in [20–28]. Other works [29–34] use the population size as an important factor to change the balance between exploration and exploitation. With a larger population size, the search space is explored more than with a smaller population size. This is an easier way to maintain the diversity, but is often an unsatisfactory solution. Without proper handling, even larger populations can converge to only one point and introduce more function evaluations. Recently, new operators have been added to several traditional evolutionary algorithms in order to improve their original exploration-exploitation capabilities. Such operators diversify the particles whenever they concentrate in a local optimum. Some methods that employ this technique can be seen in [33–38].

Either of these approaches is necessary but not sufficient to tackle the problem of the exploration–exploitation balance. Modifying the control parameters during the evolution process without incorporating new operators for improving the population diversity makes the algorithm defenseless against the premature convergence and may result in poor exploratory characteristics of the algorithm [40]. On the other hand, incorporating new operators without modifying the control parameters leads to increase the computational cost and weaken the exploitation process of candidate regions [38]. Therefore, it is reasonable to incorporate both of these approaches into a single algorithm.

In this chapter, a novel nature-inspired algorithm, namely the States of Matter Search (SMS) is presented for solving global optimization problems. The SMS algorithm is based on the simulation of the states of matter phenomenon. In SMS, individuals emulate molecules which interact to each other by using evolutionary operations based on the physical principles of the thermal energy motion mechanism. Such operations allow increasing the population diversity and avoiding the concentration of particles in a local minimum. The presented approach combines the use of the defined operators with a control strategy that modifies the parameter setting of each operation during the evolution process. Different to other approaches where the used techniques for improving the exploration–exploitation capabilities are usually added over well-kwon algorithms, SMS solves the same problem incorporating mechanisms that mimic the states of matter phenomenon. The algorithm is devised considering each state of matter one different exploration–exploitation ratio. Thus, the evolutionary process is divided into three stages which emulate the three states of matter: gas, liquid and solid. In each state, the search positions, which are modeled as molecules, exhibit different behaviors. Beginning from the gas state (pure exploration), the algorithm modifies the intensities of exploration and exploitation until the solid state (pure exploitation) is reached. As a result, the approach can substantially improve the balance between exploration–exploitation, yet preserving the good search capabilities of an EA. To illustrate the proficiency and robustness of the presented algorithm, it was compared with other well-known evolutionary methods including recent variants that incorporate diversity preservation schemas. In the comparison, it has been examined several standard benchmark functions that are usually employed within the evolutionary algorithms field. Experimental results show that the presented method achieves a good performance over its counterparts as a consequence of its better exploration–exploitation capabilities.

This chapter is organized as follows. Section 3.2 introduces basic characteristics of the three states of matter. In Sect. 3.3, the novel SMS algorithm and its characteristics are both described. Section 3.4 presents the experimental results and a comparative study. Finally, in Sect. 3.5, some conclusions are drawn.

3.2 States of the Matter

The matter can take different phases which are commonly known as states. Traditionally, three states of matter are known: solid, liquid, and gas. The differences among such states are based on the forces which are exerted among particles composing a material [41].

In the gas phase, molecules present enough kinetic energy so that the effect of intermolecular forces is small (or zero for an ideal gas), and the typical distance between neighboring molecules is greater than the molecular size. A gas has no definite shape or volume, but occupies the entire container in which it is confined. Figure 3.1a shows the movements exerted by particles in a gas state. The movement experimented by the molecules represent the maximum permissible displacement ρ_1 among particles [42]. In a liquid state, intermolecular forces are more restrictive than the gas state. The molecules have enough energy to move relatively to each other still keeping a mobile structure. Therefore, the shape of a liquid is not definite

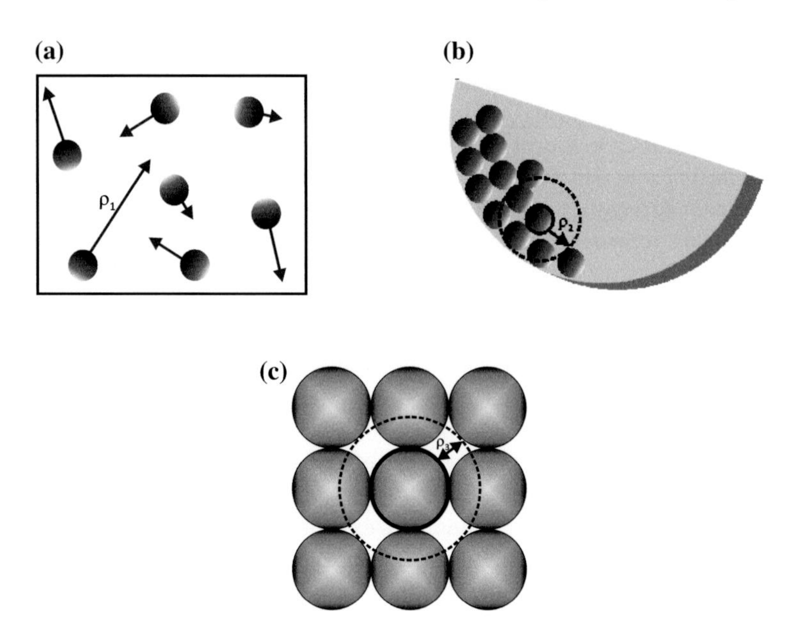

Fig. 3.1 Different states of matter: **a** gas, **b** liquid and **c** solid

but is determined by its container. Figure 3.1b presents a particle movement ρ_2 within a liquid state. Such movement is smaller than those considered by the gas state but bigger than the solid state [43]. In the solid state, the particles (or molecules) are packed together closely. The forces among particles are strong enough so that the particles cannot move freely but only vibrate. As a result, a solid has a stable, definite shape and a definite volume. Solids can only change their shape by force, as when broken or cut. Figure 3.1c shows a molecule configuration in a solid state. Under such conditions, particles are able to vibrate (be perturbed) considering a minimal ρ_3 distance [43].

In this chapter, an evolutionary algorithm is originated from the idea which relates each state matter to different exploration–exploitation relationships. Thus, the evolutionary process is divided into three phases which emulate the three states of matter: gas, liquid and solid. At each state, the individuals, who are modeled as molecules, keep different movement abilities.

3.3 States of Matter Search (SMS)

3.3.1 Definition of Operators

In the approach, agents are considered molecules whose positions on a multidimensional space are modified as the algorithm evolves. The movement of such molecules is motivated by the analogy to the thermal energy of motion.

The velocity, direction and movement of each molecule is determined considering the collision and attraction forces experimented by the set of molecules and random phenomena [44]. In our approach, such behaviors have been implemented defining several operators such as direction vector, collision and random positions. The direction vector operator assigns a direction vector to each molecule leading the particle movement. Such vector changes at each iteration as the evolution process takes place. The collision operator mimics the collisions experimented by the molecules when they interact to each other. A collision is considered when the distance between two molecules is shorter than a determined proximity distance. The collision operator is implemented by interchanging directions of the involved molecules. In order to simulate the random behavior of molecules, the presented algorithm generates random positions following a probabilistic criterion that considers random locations within the feasible search space.

The next section presents all operators that are used in the algorithm. Such operators are the same for all the states of matter; however, depending on which state is referred, they may be employed over different configurations.

3.3.1.1 Direction Vector

The direction vector operator mimics the way in which molecules change their positions as the evolution process continues. For each D-dimensional molecule $\mathbf{p}_i$ from the population P, it is assigned a D-dimensional direction vector $\mathbf{d}_i$ which stores the vector that controls the particle movement. Initially, all the direction vectors $\left(\mathbf{D} = \left\{\mathbf{d}_1, \mathbf{d}_2, \ldots, \mathbf{d}_{N_p}\right\}\right)$ are randomly set in the range of $[-1,1]$.

As the system evolves, molecules experiment several attraction forces. In order to simulate such forces, the presented algorithm implements the attraction phenomenon by moving each molecule towards the best so-far particle. Therefore, the new direction vector of each molecule is iteratively computed considering the following model:

$$\mathbf{d}_i = \mathbf{d}_i \cdot \left(1 - \frac{Ite}{gen}\right) \cdot 0.5 + \mathbf{a}_i \tag{3.1}$$

where $\mathbf{a}_i$ represents the attraction unitary vector calculated as $\mathbf{a}_i = (\mathbf{m}_{best} - \mathbf{p}_i)/\|\mathbf{m}_{best} - \mathbf{p}_i\|$, being $\mathbf{m}_{best}$ the element holding the best fitness value that it is currently stored in memory M, while $\mathbf{p}_i$ is the molecule i of the population P. *Ite* represents the iteration number whereas *gen* involves the total iteration number that constitutes the complete evolution process.

In order to calculate the new molecule position, it is necessary to compute the velocity $\mathbf{v}_i$ of each molecule by using:

$$\mathbf{v}_i = \mathbf{d}_i \cdot v_{init} \tag{3.2}$$

being v the initial velocity magnitude which is calculated as follows:

$$v_{init} = \frac{\sum_{j=1}^{N_p} \left(a_j^{high} - a_j^{low} \right)}{D} \cdot \beta \tag{3.3}$$

where a_j^{low} and a_j^{high} are the low j parameter bound and the upper j parameter bound, respectively whereas $\beta \in [0, 1]$. Then, the new position for each molecule is updated by:

$$p_i^j = p_i^j + v_i^j \cdot rand(0, 1) \cdot \rho \cdot \left(a_j^{high} - a_j^{low} \right), \quad \text{where} \quad 0.5 \leq \rho \leq 1. \tag{3.4}$$

3.3.1.2 Collision

The collision operator mimics the collisions experimented by molecules while they interact to each other. Collisions are determined in case the distance between two molecules is shorter than a determined proximity distance. Therefore, if $\|\mathbf{p}_i - \mathbf{p}_k\| < r$, a collision between molecules i and k is assumed; otherwise, there is no collision, considering $i, k \in \{1, \ldots, N_p\}$ such that $i \neq k$. If a collision occurs, the direction vector for each particle is changed by interchanging their respective direction vectors as follows:

$$\mathbf{d}_i = \mathbf{d}_k \quad \text{and} \quad \mathbf{d}_k = \mathbf{d}_i \tag{3.5}$$

The collision radius is calculated by:

$$r = \frac{\sum_{j=1}^{N_p} \left(a_j^{high} - a_j^{low} \right)}{D} \cdot \alpha, \quad \alpha \in [0, 1] \tag{3.6}$$

3.3.1.3 Random Positions

In order to simulate the random behavior of molecules, the presented algorithm generates random positions following a probabilistic criterion within the feasible search space.

For this operation, a uniform random number r_m is generated within the range [0,1]. If r_m is smaller than a threshold H, a random molecule position is generated; otherwise, the element remains with no change. Therefore such operation can be modeled as follows:

$$p_i^j = \begin{cases} a_j^{low} + rand(0,1) \cdot \left(a_j^{high} - a_j^{low}\right) & \text{with probability } H \\ p_i^j & \text{with probability } (1-H) \end{cases} \quad (3.7)$$
$$i \in \{1, \ldots, N_p\}, \quad j \in \{1, \ldots, D\}$$

3.3.1.4 Memory Update

Despite the memory update operator does not try to mimic any molecule behavior, it is used in order to store the best so-far solutions. This operator defines a memory M of the same size of population P (N_p). Only at the initial stage, the N_p molecules of P are stored in the memory **M**. In subsequent iterations, the elements of P and **M** are merged into $\mathbf{M}(\mathbf{M} = \mathbf{M} \cup \mathbf{P})$. The N_p elements holding best fitness values are preserved in $\mathbf{M}\left(\mathbf{M} = \{\mathbf{m}_1, \mathbf{m}_2, \ldots, \mathbf{m}_{N_p}\}\right)$ whereas the rest are deleted.

3.3.2 States of Matter Search (SMS) Algorithm

The algorithm is composed of three stages corresponding to the three states of matter: the gas state, the liquid state and the solid state.

Each stage has its own behavior. In the first stage, exploration is intensified whereas in the second one a mild transition between exploration and exploitation is executed. Finally, in the third phase, solutions already found are refined by emphasizing the exploitation process. The complete transition of states is organized into three different asymmetric steps, employing only 50 % of the all iterations for the gas state (exploration), 40 % for the liquid state (exploration-exploitation) and 10 % for the solid state (exploitation). The overall process is graphically described by Fig. 3.2.

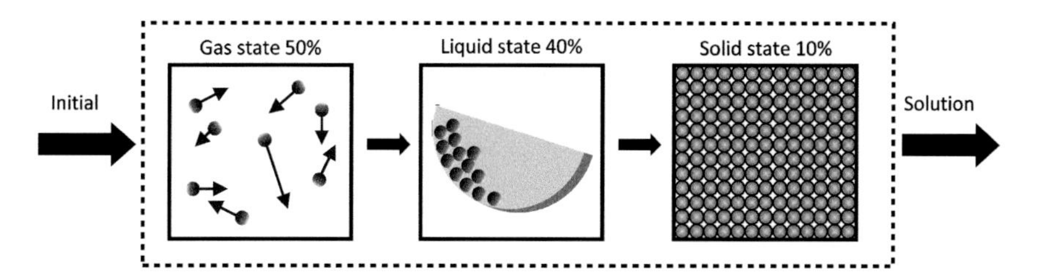

Fig. 3.2 Evolution process experimented by the presented approach

3.3.2.1 Initialization

The algorithm begins by initializing a set P of N_p molecules ($\mathbf{P} = \left\{\mathbf{p}_1, \mathbf{p}_2, \ldots, \mathbf{p}_{N_p}\right\}$). Each molecule position $\mathbf{p}_i$ is a D-dimensional vector containing the parameter values to be optimized. Such values are randomly and uniformly distributed between the pre-specified lower initial parameter bound a_j^{low} and the upper initial parameter bound a_j^{high}, just as it described by the following expressions:

$$\begin{gathered} p_i^j = a_j^{low} + rand\,(0,1) \cdot \left(a_j^{high} - a_j^{low}\right) \\ i \in \left\{1, \ldots, N_p\right\}, \quad j \in \{1, \ldots, D\} \end{gathered} \tag{3.8}$$

With j and i, being the parameter and molecule index respectively. Hence, p_i^j is the jth parameter of the ith molecule.

3.3.2.2 Gas State

In the gas state, molecules experiment severe displacements and collisions. Such state is characterized by random movements produced by non-modeled molecule phenomena [44]. Therefore, the ρ value from the direction vector operator is set to a value near to one so that the molecules can reach longer distances. Similarly, the H value representing the random positions operator is also configured to a value near to one in order to allow the random generation of more molecule positions. The computational procedure for the gas state can be summarized as follows:

Step 1	Set the parameters $\rho \in [0.8, 1]$, $\beta = 0.8$, $\alpha = 0.8$ and $H = 0.9$ being consistent with the gas state
Step 2	Obtain the new memory M using the memory update operator Sect. 3.3.1.4
Step 3	Calculate the new molecule positions P using the Direction vector operator Sect. 3.3.1.1
Step 4	Solve the collisions using the Collision operator Sect. 3.3.1.2
Step 5	Generate new random positions using the Random positions operator Sect. 3.3.1.3
Step 6	If the 50 % of the evolution process is completed, then the process continues to the liquid state procedure; otherwise go back to step 2

3.3.2.3 Liquid State

Although molecules currently at the liquid state exhibit restricted motion in comparison to the gas state, they still show a higher flexibility with respect to the solid state. Furthermore, the generation of random positions which are produced by non-modeled molecule phenomena is scarce [45]. For this reason, the ρ value from the direction vector operator is bounded to a value between 0.3 and 0.6. Similarly, the H value, from the random positions operator, is configured to a value nearby cero in order to allow the random generation of fewer molecule positions. In the liquid state, collisions are also less common than in gas state, so the collision radius controlled by α is set to a smaller value in comparison to the gas state. The computational procedure for the liquid state can be summarized as follows:

Step 7	Set the parameters $\rho \in [0.3, 0.6]$, $\beta = 0.4$, $\alpha = 0.2$ and $H = 0.2$ being consistent with the liquid state
Step 8	Obtain the new memory M using the memory update operator Sect. 3.3.1.4
Step 9	Calculate the new molecule positions P using the Direction vector operator Sect. 3.3.1.1
Step 10	Solve the collisions using the Collision operator Sect. 3.3.1.2
Step 11	Generate new random positions using the Random positions operator Sect. 3.3.1.3
Step 12	If the 90 % of the evolution process is completed, then the process continues with the solid state procedure; otherwise go back to step 8

3.3.2.4 Solid State

In the solid state, the forces among particles are strong enough so that particles cannot move freely but only vibrate. As a result, effects such as collision and generation of random positions are not considered in the solid state [44]. Therefore, the ρ value of the direction vector operator is set to a value nearby zero indicating that the molecules can only vibrate around their original positions. The computational procedure for the solid state can be summarized as follows:

Step 13	Set the parameters $\rho \in [0.0, 0.1]$ and $\beta = 0.1$, being consistent with the solid state
Step 14	Obtain the new memory M using the Update memory operator Sect. 3.3.1.4
Step 15	Calculate the new molecule positions P using the Direction vector operator Sect. 3.3.1.1
Step 16	If the 100 % of the evolution process is completed, the process is finished; otherwise go back to step 14

3.4 Experimental Results

A comprehensive set of 14 functions, collected from Refs. [46–50], has been used to test the performance of the presented approach. Tables 3.1, 3.2 and 3.3 present the benchmark functions used in our experimental study. Such functions are

Table 3.1 Unimodal test functions

Test function	S	f_{opt}	n
$f_1(x) = \sum_{i=1}^{n} x_i^2$	$[-100, 100]^n$	0	30
$f_2(x) = \max\{\lvert x_i \rvert, 1 \le i \le n\}$	$[-100, 100]^n$	0	30
$f_3(x) = \sum_{i=1}^{n-1} \left[100(x_{i+1} x_i^2)^2 + (x_i - 1)^2\right]$	$[-30, 30]^n$	0	30
$f_4(x) = \sum_{i=1}^{n} i x_i^4 + rand(0, 1)$	$[-1.28, 1.28]^n$	0	30

classified into three different categories: Unimodal test functions (Table 3.1), multimodal test functions (Table 3.2) and multimodal test functions with fixed dimensions (Table 3.3). In such tables, n indicates the dimension of the function, f_{opt} the optimum value of the function and S the subset of R^n. The functions optimum position (x_{opt}) for $f_1,, f_2, f_4, f_6, f_7, f_{10}, f_{11}$ and f_{14} is at $x_{opt} = [0]^n$, for f_3, f_8 and f_9 is at $x_{opt} = [1]^n$, for f_5 is at $x_{opt} = [420.96]^n$, for f_{18} is at $x_{opt} = [0]^n$, for f_{12} is at $x_{opt} = [0.0003075]^n$ and for f_{13} is at $x_{opt} = [-3.32]^n$. A detailed description of optimum locations is given in each table.

3.4.1 Performance Comparison to Other Metaheuristic Algorithms

The MSM algorithm was tested over 14 functions comparing the results with those produced by the Gravitational Search Algorithm (GSA) [15], Particle Swarm Optimization (PSO) [16] and Differential Evolution (DE) [12]. In all cases, the population is set to 50. The maximum iteration number for functions in Tables 3.1 and 3.2 is set to 1000 and for functions in Table 3.3 is set to 500. This stop criterion is selected to maintain compatibility to similar works reported in the literature [3, 15].

The parameter setting for each algorithm in the comparison is described as follows:

1. GSA [15]: the parameters are set as $G_o = 100$ and $\alpha = 20$; the total number of iterations is set to 1000 for functions f_1 to f_{11} and 500 for functions f_{12} to f_{14}. The total number of agents is set to 50. Such values are the best parameter set for this algorithm according to [15].
2. PSO [16]: the parameters are set as $c_1 = 2$ and $c_2 = 2$, besides the weight factor decreases linearly from 0.9 to 0.2.
3. DE [12]: the DE/Rand/1 scheme is employed. The crossover probability is set to $CR = 0.9$ and the weight factor is set to $F = 0.8$.

The experimental comparison among metaheuristic algorithms with respect to SMS is developed according to the function type as follows:

Table 3.2 Multimodal test functions

Test function	S	f_{opt}	n
$f_5(x) = 418.9829n + \sum_{i=1}^{n} \left(-x_i \sin\left(\sqrt{\lvert x_i \rvert}\right)\right)$	$[-500, 500]^n$	0	30
$f_6(x) = \sum_{i=1}^{50} \left(x_i^2 - 10\cos(2\pi x_i) + 10\right)$	$[-5.12, 5.12]^n$	0	30
$f_7(x) = \frac{1}{4000}\sum_{i=1}^{n} x_i^2 - \prod_{i=1}^{n} \cos\left(\frac{x_i}{\sqrt{i}}\right) + 1$	$[-600, 600]^n$	0	30
$f_8(x) = \frac{\pi}{n}\left\{10\sin(\pi y_1) + \sum_{i=1}^{n-1} (y_i - 1)^2\left[1 + 10\sin^2(\pi y_{i+1})\right] + \left(y_n - 1^2\right)\right\}\ldots$ $+ \sum_{i=1}^{n} u(x_i, 10, 100, 4)$ $y_i = 1 + \frac{(x_i+1)}{4} \quad u(x_i, a, k, m) = \begin{cases} k(x_i - a)^m & x_i > a \\ 0 & -a \le x_i \le a \\ k(-x_i - a)^m & x_i < a \end{cases}$	$[-50, 50]^n$	0	30
$f_9(x) = 0.1\left\{\sin^2(3\pi x_1) + \sum_{i=1}^{n} (x_i - 1)^2\left[1 + \sin^2(3\pi x_i + 1)\right]\ldots\right.$ $\left. + (x_n - 1)^2\left[1 + \sin^2(2\pi x_n)\right]\right\} + \sum_{i=1}^{n} u(x_i, 5, 100, 4)$ where $u(x_i, a, k, m)$ is the same as f_8	$[-50, 50]^n$	0	30
$f_{10}(x) = \sum_{i=1}^{n} x_i^2 + \left(\sum_{i=1}^{n} 0.5ix_i\right)^2 + \left(\sum_{i=1}^{n} 0.5ix_i\right)^4$	$[-10, 10]^n$	0	30
$f_{11}(x) = 1 - \cos(2\pi\lVert x\rVert) + 0.1\lVert x\rVert$ where $\lVert x\rVert = \sqrt{\sum_{i=1}^{n} x_i^2}$	$[-100, 100]^n$	0	30

Table 3.3 Multimodal test functions with fixed dimensions

Test function	S	f_{opt}	n
$f_{12}(x) = \sum_{i=1}^{11} \left[a_i - \frac{x_i\left(b_i^2 + b_i x_2\right)}{b_i^2 + b_i x_3 + x_4} \right]^2$ $\mathbf{a} = [0.1957, 0.1947, 0.1600, 0.0844, 0.0627, 0.456, 0.0342, 0.0323, 0.0235, 0.0246]$ $\mathbf{b} = [0.25, 0.5, 1, 2, 4, 6, 8, 10, 12, 14, 16]$	$[-5, 5]^n$	0.00030	4
$f_{13}(x) = \sum_{i=1}^{4} c_i \exp\left(-\sum_{j=1}^{3} A_{ij}\left(x_j - P_{ij}\right)^2\right)$ $\mathbf{A} = \begin{bmatrix} 10 & 3 & 17 & 3.5 & 1.7 & 8 \\ 0.05 & 10 & 17 & 0.1 & 8 & 14 \\ 3 & 3.5 & 17 & 10 & 17 & 8 \\ 17 & 8 & 0.05 & 10 & 0.1 & 14 \end{bmatrix}$ $\mathbf{c} = [1, 1.2, 3, 3.2]$ $\mathbf{P} = \begin{bmatrix} 0.131 & 0.169 & 0.556 & 0.012 & 0.828 & 0.588 \\ 0.232 & 0.413 & 0.830 & 0.373 & 0.100 & 0.999 \\ 0.234 & 0.141 & 0.352 & 0.288 & 0.304 & 0.665 \\ 0.404 & 0.882 & 0.873 & 0.574 & 0.109 & 0.038 \end{bmatrix}$	$[0, 1]^n$	−3.32	6
$f_{14}(x) = (1.5 - x_1(1 - x_2))^2 + (2.25 - x_1(1 - x_2))^2 + (2.625 - x_1(1 - x_2))^2$	$[-4.5, 4.5]^n$	0	2

(a) Unimodal test functions (Table 3.1).
(b) Multimodal test functions (Table 3.2).
(c) Multimodal test functions with fixed dimension (Table 3.3).

(a) *Unimodal test functions*

This experiment is performed over the functions presented in Table 3.1. Such test compares the SMS against the algorithms GSA, PSO and DE. The results for 30 runs are reported in Table 3.1 considering the following performance indexes: the Average Best-so-far (AB) solution, the Median Best-so-far (MB) and the Standard Deviation (SD) of best-so-far solution. The best outcome for each function is boldfaced. According to this table, SMS provides better results than GSA, PSO and DE for all functions. In particular, the test yields the largest difference in performance which is directly related to a better trade-off between exploration and exploitation. Just as it is illustrated in Fig. 3.3, SMS and GSA have similar convergence rates at finding the optimal minimal, yet faster than PSO and DE.

A non-parametric statistical significance proof known as the Wilcoxon's rank sum test for independent samples [51, 52] has been conducted with an 5 % significance level, over the "average best-so-far" (AB) data of Table 3.4. Table 3.5 reports the p-values produced by Wilcoxon's test for the pair-wise comparison of the "average best so-far" of four groups. Such groups are formed by SMS versus GSA, SMS versus PSO and SMS versus DE. As a null hypothesis, it is assumed that there is no significant difference between mean values of the two algorithms. The alternative hypothesis considers a significant difference between the "average best-so-far" values of both approaches. All p-values reported in Table 3.5 are less than 0.05 (5 % significance level) which is a strong evidence against the null

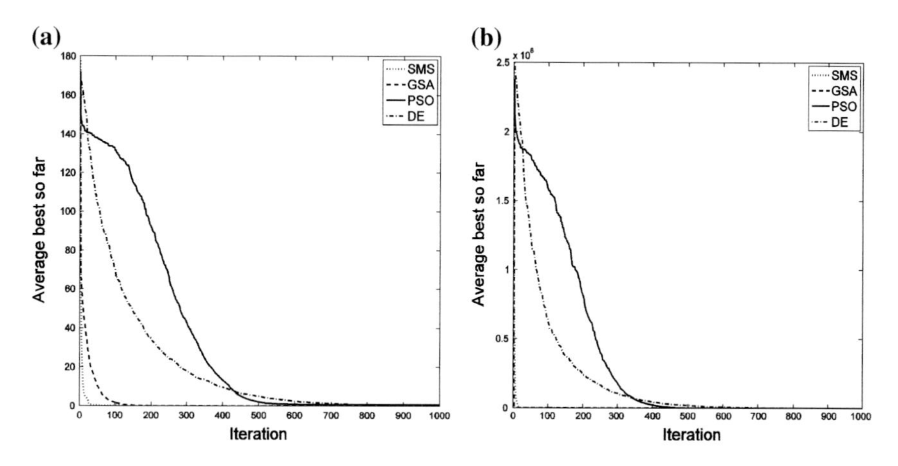

Fig. 3.3 Convergence rate comparison of GSA, PSO, DE and SMS for minimization of **a** f_1 and **b** f_3 considering $n = 30$

Table 3.4 Minimization result of benchmark functions of Table 3.1 with $n = 30$

		SMS	GSA	PSO	DE
$f_1(x)$	AB	4.68457E−16	1.3296E−05	0.873813333	0.1865842
	MB	4.50542E−16	7.46803E−06	4.48139E−12	0.189737658
	SD	1.23694E−16	1.45053E−05	4.705628811	0.039609704
$f_2(x)$	AB	0.033116745	0.173618066	12.83021186	54.85755486
	MB	1.02069E−08	0.1599328	12.48059177	54.59915941
	SD	0.089017369	0.122230612	3.633980625	4.506836836
$f_3(x)$	AB	19.64056183	32.83253962	33,399.69716	46,898.34558
	MB	26.87914282	27.65055745	565.0810149	43,772.19502
	SD	11.8115879	19.11361524	43,099.34439	15,697.6366
$f_4(x)$	AB	8.882513655	9.083435186	5.05362961	12.83391861
	MB	9.016816582	9.150769929	13.91301428	12.89762202
	SD	0.442124359	0.499181789	4.790792877	0.542197802

Maximum number of iterations = 1000

Table 3.5 p-values produced by Wilcoxon's test comparing SMS versus PSO, SMS versus GSA and SMS versus DE over the "average best-so-far" (AB) values from Table 3.4

SMS versus	PSO	GSA	DE
f_1	3.94×10^{-5}	7.39×10^{-4}	1.04×10^{-6}
f_2	5.62×10^{-5}	4.92×10^{-4}	2.21×10^{-6}
f_3	6.42×10^{-8}	7.11×10^{-7}	1.02×10^{-4}
f_4	1.91×10^{-8}	7.39×10^{-4}	1.27×10^{-6}

hypothesis. Therefore, such evidence indicates that the SMS results are statistically significant and that it has not occurred by coincidence (i.e. due to the normal noise contained in the process).

(b) *Multimodal test functions*

Multimodal functions represent a good optimization challenge as they possess many local minima (Table 3.2). In the case of multimodal functions, final results are very important since they reflect the algorithm's ability to escape from poor local optima and locate a near-global optimum. There are developed experiments on f_5 to f_{11} where the number of local minima increases exponentially as the dimension of the function increases. The dimension of these functions is set to 30. The results are averaged over 30 runs, reporting the performance indexes in Table 3.6 as follows: the Average Best-so-far (AB) solution, the Median Best-so-far (MB) and the Standard Deviation (SD) best-so-far (the best result for each function is highlighted). Likewise, p-values of the Wilcoxon signed-rank test of 30 independent runs are listed in Table 3.7.

For f_8, f_9, f_{10} and f_{11}, SMS yields a much better solutions than other methods. However, for functions f_5, f_6 and f_7, SMS produces similar results to GSA. The

Table 3.6 Minimization result of benchmark functions in Table 3.2 with $n = 30$

		SMS	GSA	PSO	DE
$f_5(x)$	AB	1756.862345	9750.440145	4329.650468	4963.600685
	MB	0.070624076	9838.388135	4233.282929	5000.245932
	SD	1949.048601	405.1365297	699.7276454	202.2888921
$f_6(x)$	AB	10.95067665	15.18970458	130.5959941	194.6220253
	MB	0.007142491	13.9294268	129.4942809	196.1369499
	SD	14.38387472	4.508037915	27.87011038	9.659933059
$f_7(x)$	AB	0.000299553	0.000575111	0.19630233	0.98547042
	MB	8.67349E−05	0	0.011090373	0.991214493
	SD	0.000623992	0.0021752	0.702516846	0.031985616
$f_8(x)$	AB	1.35139E−05	2.792846799	145,066,676.9	304.6986718
	MB	7.14593E−06	2.723230534	0.675050254	51.86661185
	SD	2.0728E−05	1.324814757	170,879,878.5	554.2231579
$f_9(x)$	AB	0.002080591	14.49783478	13,668,886.94	67251.29956
	MB	0.000675275	9.358377669	7.00288E−05	37,143.43153
	SD	0.003150999	18.02351657	73,609,207.58	63,187.52749
$f_{10}(x)$	AB	0.003412411	40.59204902	365.7806149	822.087914
	MB	0.003164797	39.73690704	359.104488	829.1521586
	SD	0.001997493	11.46284891	148.9342039	81.93476435
$f_{11}(x)$	AB	0.199873346	1.121397135	0.857971914	3.703467688
	MB	0.199873346	1.114194975	0.499967033	3.729096071
	SD	0.073029674	0.271747312	1.736399225	0.278860779

Maximum number of iterations = 1000

Table 3.7 p-values produced by Wilcoxon's test comparing SMS versus GSA, SMS versus PSO and SMS versus DE over the "average best-so-far" (AB) values from Table 3.6

SMS versus	GSA	SO	DE
f_5	0.087	8.38×10^{-4}	4.61×10^{-4}
f_6	0.062	1.92×10^{-9}	9.97×10^{-8}
f_7	0.055	4.21×10^{-5}	3.34×10^{-4}
f_8	7.74×10^{-9}	3.68×10^{-7}	8.12×10^{-5}
f_9	1.12×10^{-8}	8.80×10^{-9}	4.02×10^{-8}
f_{10}	4.72×10^{-9}	3.92×10^{-5}	2.20×10^{-4}
f_{11}	4.72×10^{-9}	3.92×10^{-5}	2.20×10^{-4}

Wilcoxon rank test results, presented in Table 3.7, demonstrate that SMS performed better than RGA, PSO, GSA and DE considering the four problems $f_8 - f_{11}$, whereas, from a statistical viewpoint, there is no difference between results from SMS and GSA for f_5, f_6 and f_7. The progress of the "average best-so-far" solution over 30 runs for functions f_5 and f_{11} are shown in Fig. 3.4.

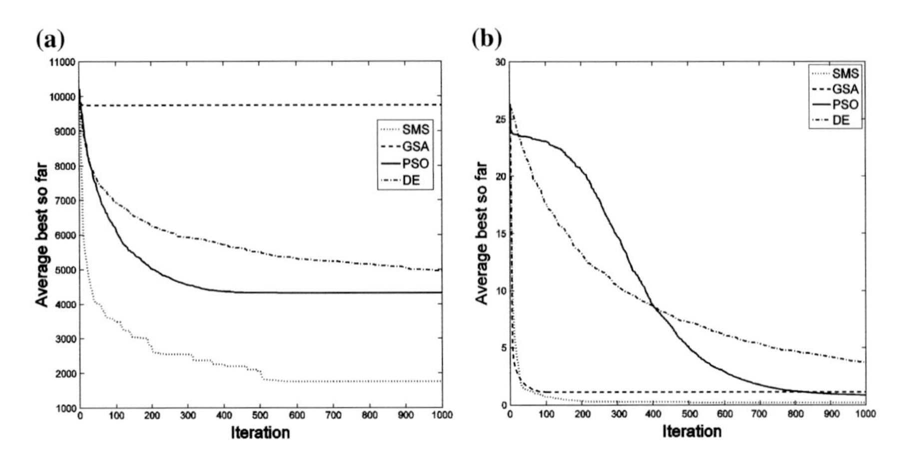

Fig. 3.4 Convergence rate comparison of RGA, PSO, GSA, DE and SMS for minimization of **a** f_5 and **b** f_{11} considering $n = 30$

Table 3.8 Minimization results of benchmark functions in Table 3.3 with $n = 30$

		SMS	GSA	PSO	DE
$f_{12}(x)$	AB	0.004361206	0.051274735	0.020521847	0.006247895
	MB	0.004419241	0.051059414	0.020803912	0.004361206
	SD	0.004078875	0.016617355	0.021677285	8.7338E−15
$f_{13}(x)$	AB	−3.862782148	−3.207627571	−3.122812884	−3.200286885
	MB	−3.862782148	−3.222983851	−3.198877457	−3.200286885
	SD	2.40793E−15	0.032397257	0.357126056	2.22045E−15
$f_{14}(x)$	AB	0	0.00060678	1.07786E−11	4.45378E−31
	MB	3.82624E−12	0.000606077	0	4.93038E−32
	SD	2.93547E−11	0.000179458	0	1.0696E−30

Maximum number of iterations = 500

(c) *Multimodal test functions with fixed dimensions*

In the following experiments, the SMS algorithm is compared with GSA, PSO and DE over a multidimensional functions set with fixed dimensions widely used in the meta-heuristic literature. The functions used for the experiments are f_{12}, f_{13} and f_{14} which are presented in Table 3.3. The results, reported in Table 3.8, show that GSA, PSO and DE have similar values in their performance. The evidence shows how meta-heuristic algorithms maintain a similar average performance when they faced low-dimensional functions [53]. Figure 3.5 presents the convergence rate for the GSA, PSO, DE and SMS algorithms considering functions f_{12} to f_{13}.

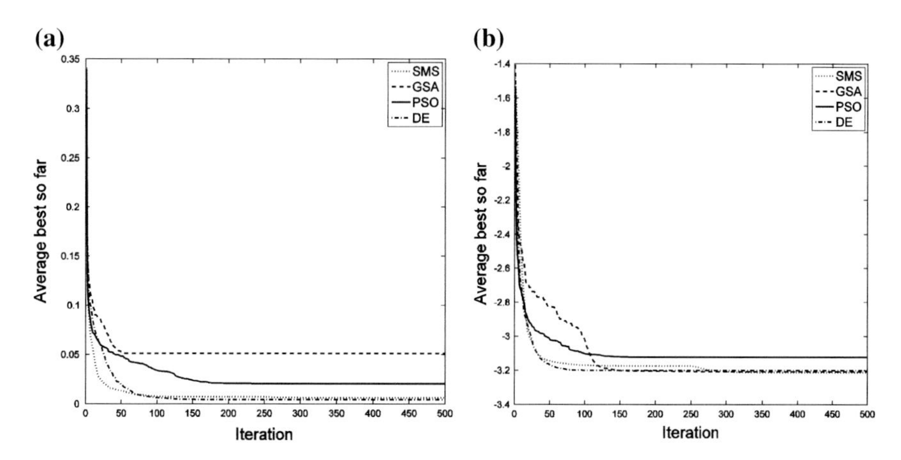

Fig. 3.5 Convergence rate comparison of PSO, GSA, DE and SMS for minimization of **a** f_{12} and **b** f_{13}

3.5 Summary

In this chapter, an evolutionary algorithm for global optimization called the States of Matter Search (SMS) has been introduced. The algorithm is originated from relating each state of matter to one different exploration–exploitation relationship. Thus, the evolutionary process is divided into three phases which emulate the three states of matter: gas, liquid and solid. In each state, the individuals, who are modeled as molecules, maintain different movement capacities. Beginning from the gas state (pure exploration), the algorithm modifies the intensities of exploration and exploitation until the solid state (pure exploitation) is reached. As a result, the approach can substantially improve the balance between exploration–exploitation, yet preserving the good search capabilities of an EA approach.

SMS has been experimentally tested considering a suite of 14 benchmark functions. The performance of SMS has been also compared with the following metaheuristic algorithms: the Particle Swarm Optimization (PSO) [16], the Gravitational Search Algorithm (GSA) [15] and the Differential Evolution (DE) method [12]. The experiments have demonstrated that SMS outperforms other metaheuristic algorithms for unimodal and multimodal benchmark functions regarding their solution quality while keeping a performance similar to other EA when it faced low-dimensional functions. Such result is congruent to other works reported in the literature according to [53].

The SMS's remarkable performance is associated with two different reasons: (i) the defined operators allow a better particle distribution in the search space, increasing the algorithm's ability to find the global optima; and (ii) the division of the evolution process, at different states, provides different rates between exploration and exploitation during the evolution process. At the beginning, pure exploration is favored at the gas state, then a mild transition between exploration and exploitation features during liquid state and finally, during the solid state, pure exploitation is performed.

References

1. Pardalos Panos, M., Romeijn Edwin, H., Toy, H.: Recent developments and trends in global optimization. J. Comput. Appl. Math. **124**, 209–228 (2000)
2. Floudas, C., Akrotirianakis, I., Caratzoulas, S., Meyer, C., Kallrath, J.: Global optimization in the 21st century: advances and challenges. Comput. Chem. Eng. **29**(6), 1185–1202 (2005)
3. Ying, J., Ke-Cun, Z., Shao-Jian, Q.: A deterministic global optimization algorithm. Appl. Math. Comput. **185**(1), 382–387 (2007)
4. Georgieva, A., Jordanov, I.: Global optimization based on novel heuristics, low-discrepancy sequences and genetic algorithms. Eur. J. Oper. Res. **196**, 413–422 (2009)
5. Lera, D., Sergeyev, Y.: Lipschitz and Hölder global optimization using space-filling curves. Appl. Numer. Math. **60**(1–2), 115–129 (2010)
6. Fogel, L.J., Owens, A.J., Walsh, M.J.: Artificial Intelligence Through Simulated Evolution. Wiley, Chichester, UK (1966)
7. De Jong, K.: Analysis of the behavior of a class of genetic adaptive systems. Ph.D. thesis, University of Michigan, Ann Arbor, MI (1975)
8. Koza, J.R.: Genetic programming: a paradigm for genetically breeding populations of computer programs to solve problems. Rep. No. STAN-CS-90-1314, Stanford University, CA (1990)
9. Holland, J.H.: Adaptation in Natural and Artificial Systems. University of Michigan Press, Ann Arbor, MI (1975)
10. Goldberg, D.E.: Genetic Algorithms in Search, Optimization and Machine Learning. Addison Wesley, Boston, MA (1989)
11. de Castro, L.N., Von Zuben, F.J.: Artificial immune systems: Part I—basic theory and applications. Technical report, TR-DCA 01/99. December 1999
12. Storn, R., Price, K.: Differential evolution-a simple and efficient adaptive scheme for global optimisation over continuous spaces. Tech. Rep. TR-95-012. ICSI, Berkeley, Calif (1995)
13. Kirkpatrick, S., Gelatt, C., Vecchi, M.: Optimization by simulated annealing. Science **220** (4598), 671–680 (1983)
14. İlker, B., Birbil, S., Shu-Cherng, F.: An electromagnetism-like mechanism for global optimization. J. Global Optim. **25**, 263–282 (2003)
15. Rashedia, E., Nezamabadi-pour, H., Saryazdi, S.: Filter modeling using gravitational search algorithm. Eng. Appl. Artif. Intell. **24**(1), 117–122 (2011)
16. Kennedy, J., Eberhart, R.: Particle swarm optimization. In: Proceedings of the 1995 IEEE International Conference on Neural Networks, vol. 4, pp. 1942–1948, December 1995
17. Dorigo, M., Maniezzo, V., Colorni, A.: Positive feedback as a search strategy. Technical Report No. 91-016, Politecnico di Milano (1991)
18. Tan, K.C., Chiam, S.C., Mamun, A.A., Goh, C.K.: Balancing exploration and exploitation with adaptive variation for evolutionary multi-objective optimization. Eur. J. Oper. Res. **197**, 701–713 (2009)
19. Chen, G., Low, C.P., Yang, Z.: Preserving and exploiting genetic diversity in evolutionary programming algorithms. IEEE Trans. Evol. Comput. **13**(3), 661–673 (2009)
20. Liu, S.-H., Mernik, M., Bryant, B.: To explore or to exploit: an entropy-driven approach for evolutionary algorithms. Int. J. Knowl. Based Intell. Eng. Syst. **13**(3), 185–206 (2009)
21. Alba, E., Dorronsoro, B.: The exploration/exploitation tradeoff in dynamic cellular genetic algorithms. IEEE Trans. Evol. Comput. **9**(3), 126–142 (2005)

22. Fister, I., Mernik, M., Filipič, B.: A hybrid self-adaptive evolutionary algorithm for marker optimization in the clothing industry. Appl. Soft Comput. **10**(2), 409–422 (2010)
23. Gong, W., Cai, Z., Jiang, L.: Enhancing the performance of differential evolution using orthogonal design method. Appl. Math. Comput. **206**(1), 56–69 (2008)
24. Joan-Arinyo, R., Luzon, M.V., Yeguas, E.: Parameter tuning of pbil and chc evolutionary algorithms applied to solve the root identification problem. Appl. Soft Comput. **11**(1), 754–767 (2011)
25. Mallipeddi, R., Suganthan, P.N., Pan, Q.K., Tasgetiren, M.F.: Differential evolution algorithm with ensemble of parameters and mutation strategies. Appl. Soft Comput. **11**(2), 1679–1696 (2011)
26. Sadegh, M., Reza, M., Palhang, M.: LADPSO: using fuzzy logic to conduct PSO algorithm. Appl. Intell. **37**(2), 290–304 (1012)
27. Yadav, P., Kumar, R., Panda, S.K., Chang, C.S.: An intelligent tuned harmony search algorithm for optimization. Inf. Sci. **196**(1), 47–72 (2012)
28. Khajehzadeh, M., Taha, M.R., El-Shafie, A., Eslami, M.: A modified gravitational search algorithm for slope stability analysis. Eng. Appl. Artif. Intell. **25**(8), 1589–1597 (2012)
29. Koumousis, V., Katsaras, C.P.: A saw-tooth genetic algorithm combining the effects of variable population size and reinitialization to enhance performance. IEEE Trans. Evol. Comput. **10**(1), 19–28 (2006)
30. Han, Ming-Feng, Liao, Shih-Hui, Chang, Jyh-Yeong, Lin, Chin-Teng: Dynamic group-based differential evolution using a self-adaptive strategy for global optimization problems. Appl. Intell. (2012). doi:10.1007/s10489-012-0393-5
31. Brest, J., Maučec, M.S.: Population size reduction for the differential evolution algorithm. Appl. Intell. **29**(3), 228–247 (2008)
32. Li, Y., Zeng, X.: Multi-population co-genetic algorithm with double chain-like agents structure for parallel global numerical optimization. Appl. Intell. **32**(3), 292–310 (2010)
33. Paenke, I., Jin, Y., Branke, J.: Balancing population- and individual-level adaptation in changing environments. Adapt. Behav. **17**(2), 153–174 (2009)
34. Araujo, L., Merelo, J.J.: Diversity through multiculturality: assessing migrant choice policies in an island model. IEEE Trans. Evol. Comput. **15**(4), 456–468 (2011)
35. Gao, H., Xu, W.: Particle swarm algorithm with hybrid mutation strategy. Appl. Soft Comput. **11**(8), 5129–5142 (2011)
36. Jia, D., Zheng, G., Khan, M.K.: An effective memetic differential evolution algorithm based on chaotic local search. Inf. Sci. **181**(15), 3175–3187 (2011)
37. Lozano, M., Herrera, F., Cano, J.R.: Replacement strategies to preserve useful diversity in steady-state genetic algorithms. Inf. Sci. **178**(23), 4421–4433 (2008)
38. Ostadmohammadi, B., Mirzabeygi, P., Panahi, M.: An improved PSO algorithm with a territorial diversity-preserving scheme and enhanced exploration–exploitation balance. Swarm and Evolutionary Computation (in press)
39. Adra, S.F., Fleming, P.J.: Diversity management in evolutionary many-objective optimization. IEEE Trans. Evol. Comput. **15**(2), 183–195 (2011)
40. Črepineš, M., Liu, S.H., Mernik, M.: Exploration and exploitation in evolutionary algorithms: a survey. ACM Comput. Surv. **1**(1), 1–33 (2011)
41. Ceruti, G., Rubin, H.: Infodynamics: analogical analysis of states of matter and information. Inf. Sci. **177**, 969–987 (2007)
42. Chowdhury, D., Stauffer, D.: Principles of Equilibrium Statistical Mechanics, 1 edn. Wiley-VCH, 2000
43. Betts, D.S., Roy, E.: Turner Introductory Statistical Mechanics, 1 edn. Addison Wesley, 1992
44. Cengel, Y.A., Boles, M.A.: Thermodynamics: An Engineering Approach, 5th edn. McGraw-Hill, 2005
45. Bueche, F., Hecht, E.: Schaum's Outline of College Physics, 11th edn. McGraw-Hill, 2012
46. Piotrowski, A.P., Napiorkowski, J.J., Kiczko, A.: Differential evolution algorithm with separated groups for multi-dimensional optimization problems. Eur. J. Oper. Res. **216**(1), 33–46 (2012)

47. Cocco Mariani, V., Justi Luvizotto, L.G., Alessandro Guerra, F., dos Santos Coelho, L.: A hybrid shuffled complex evolution approach based on differential evolution for unconstrained optimization. Appl. Math. Comput. **217**(12), 5822–5829 (2011)
48. Yao, X., Liu, Y., Lin, G.: Evolutionary programming made faster. IEEE Trans. Evol. Comput. **3**(2), 82–102 (1999)
49. Moré, J.J., Garbow, B.S., Hillstrom, K.E.: Testing unconstrained optimization software. ACM Trans. Math. Softw. **7**(1), 17–41 (1981)
50. Tsoulos, I.G.: Modifications of real code genetic algorithm for global optimization. Appl. Math. Comput. **203**(2), 598–607 (2008)
51. Wilcoxon, F.: Individual comparisons by ranking methods. Biometrics **1**, 80–83 (1945)
52. Garcia, S., Molina, D., Lozano, M., Herrera, F.: A study on the use of non-parametric tests for analyzing the evolutionary algorithms' behaviour: a case study on the CEC'2005 Special session on real parameter optimization. J. Heurist. (2008). doi:10.1007/s10732-008-9080-4
53. Shilane, D., Martikainen, J., Dudoit, S., Ovaska, S.: A general framework for statistical performance comparison of evolutionary computation algorithms. Inf. Sci. **178**, 2870–2879 (2008)

Chapter 4
The Collective Animal Behavior method

4.1 Introduction

Global Optimization (GO) has yielded remarkable applications to many areas of science, engineering, economics and others through mathematical modelling [1]. In general, the goal is to find a global optimum of an objective function which has been defined over a given search space. Global optimization algorithms are usually broadly divided into deterministic and metaheuristic [2]. Since deterministic methods only provide a theoretical guarantee of locating a local minimum of the objective function, they often face great difficulties in solving global optimization problems [3]. On the other hand, metaheuristic methods are usually faster in locating a global optimum [4]. They virtuously adapt to black-box formulations and extremely ill-behaved functions, whereas deterministic methods usually require some theoretical assumptions about the problem formulation and its analytical properties (such as Lipschitz continuity) [5].

Several metaheuristic algorithms have been developed by a combination of rules and randomness mimicking several phenomena. Such phenomena include evolutionary processes e.g. the evolutionary algorithm proposed by Fogel et al. [6], De Jong [7], and Koza [8], the genetic algorithm (GA) proposed by Holland [9] and Goldberg [10] and the artificial immune systems proposed by De Castro et al. [11]. On the other hand, physical processes consider the simulated annealing proposed by Kirkpatrick et al. [12], the electromagnetism-like algorithm proposed by İlker et al. [13], the gravitational search algorithm proposed by Rashedi et al. [14] and the musical process of searching for a perfect state of harmony, which has been proposed by Geem et al. [15], Lee and Geem [16] and Geem [17].

Many studies have been inspired by animal behavior phenomena for developing optimization techniques. For instance, the Particle swarm optimization (PSO) algorithm which models the social behavior of bird flocking or fish schooling [18]. PSO consists of a swarm of particles which move towards best positions, seen so far, within a searchable space of possible solutions. Another behavior-inspired

E. Cuevas et al., *Advances of Evolutionary Computation: Methods and Operators*, Studies in Computational Intelligence 629,
DOI 10.1007/978-3-319-28503-0_4

approach is the Ant Colony Optimization (ACO) algorithm proposed by Dorigo et al. [19], which simulates the behavior of real ant colonies. Main features of the ACO algorithm are the distributed computation, the positive feedback and the constructive greedy search. Recently, a new metaheuristic approach which is based on the animal behavior while hunting has been proposed in [20]. Such algorithm considers hunters as search positions and preys as potential solutions.

Just recently, the concept of individual-organization [21, 22] has been widely referenced to understand collective behavior of animals. The central principle of individual-organization is that simple repeating interactions between individuals can produce complex behavioral patterns at group level [21, 23, 24]. Such inspiration comes from behavioral patterns previously seen in several animal groups. Examples include ant pheromone trail networks, aggregation of cockroaches and the migration of fish schools, all of which can be accurately described in terms of individuals following simple sets of rules [25]. Some examples of these rules [24, 26] are keeping the current position (or location) for best individuals, local attraction or repulsion, random movements and competition for the space within of a determined distance.

On the other hand, new studies [27–29] have also shown the existence of collective memory in animal groups. The presence of such memory establishes that the previous history of the group structure influences the collective behavior exhibited in future stages. According to such principle, it is possible to model complex collective behaviors by using simple individual rules and configuring a general memory.

In this chapter, a new optimization algorithm inspired by the collective animal behavior is presented. In this algorithm, the searcher agents emulate a group of animals that interact to each other based on simple behavioral rules which are modeled as mathematical operators. Such operations are applied to each agent considering that the complete group has a memory storing their own best positions seen so far, by using a competition principle. The presented approach has been compared to other well-known optimization methods. The results confirm a high performance of the presented method for solving various benchmark functions.

This chapter is organized as follows. In Sect. 4.2, we introduce the basic biologic aspects of the algorithm. In Sect. 4.3, the novel CAB algorithm and its characteristics are both described. Section 4.4 presents the experimental results and the comparative study. Finally, in Sect. 4.5, the conclusions are discussed.

4.2 Biological Fundamentals

The remarkable collective behavior of organisms such as swarming ants, schooling fish and flocking birds has long captivated the attention of naturalists and scientists. Despite a long history of scientific research, the relationship between individuals and group-level properties has just recently begun to be deciphered [30].

Grouping individuals often have to make rapid decisions about where to move or what behavior to perform in uncertain and dangerous environments. However, each individual typically has only a relatively local sensing ability [31]. Groups are, therefore, often composed of individuals that differ with respect to their informational status and individuals are usually not aware of the informational state of others [32], such as whether they are knowledgeable about a pertinent resource or about a threat.

Animal groups are based on a hierarchic structure [33] which considers different individuals according to a fitness principle called Dominance [34] which is the domain of some individuals within a group that occurs when competition for resources leads to confrontation. Several studies [35, 36] have found that such animal behavior lead to more stable groups with better cohesion properties among individuals.

Recent studies have begun to elucidate how repeated interactions among grouping animals scale to collective behavior. They have remarkably revealed that collective decision-making mechanisms across a wide range of animal group types, from insects to birds (and even among humans in certain circumstances) seem to share similar functional characteristics [21, 25, 37]. Furthermore, at a certain level of description, collective decision-making by organisms shares essential common features such as a general memory. Although some differences may arise, there are good reasons to increase communication between researchers working in collective animal behavior and those involved in cognitive science [24].

Despite the variety of behaviors and motions of animal groups, it is possible that many of the different collective behavioral patterns are generated by simple rules followed by individual group members. Some authors have developed different models, one of them, known as the self-propelled particle (SPP) model, attempts to capture the collective behavior of animal groups in terms of interactions between group members which follow a diffusion process [38–41].

On other hand, following a biological approach, Couzin et al. [24, 25] have proposed a model in which individual animals follow simple rules of thumb: (1) keep the current position (or location) for best individuals; (2) move from or to nearby neighbors (local attraction or repulsion); (3) move randomly and (4) compete for the space within of a determined distance. Each individual thus admits three different movements: attraction, repulsion or random and holds two kinds of states: preserve the position or compete for a determined position. In the model, the movement, which is executed by each individual, is decided randomly (according to an internal motivation). On the other hand, the states follow a fixed criteria set.

The dynamical spatial structure of an animal group can be explained in terms of its history [36]. Despite such a fact, the majority of studies have failed in considering the existence of memory in behavioral models. However, recent research [27, 42] have also shown the existence of collective memory in animal groups. The presence of such memory establishes that the previous history of the group structure influences the collective behavior which is exhibited in future stages. Such memory can contain the location of special group members (the dominant individuals) or the averaged movements produced by the group.

According to these new developments, it is possible to model complex collective behaviors by using simple individual rules and setting a general memory. In this work, the behavioral model of animal groups inspires the definition of novel evolutionary operators which outline the CAB algorithm. A memory is incorporated to store best animal positions (best solutions) considering a competition-dominance mechanism.

4.3 Collective Animal Behavior Algorithm (CAB)

The CAB algorithm assumes the existence of a set of operations that resembles the interaction rules that model the collective animal behavior. In the approach, each solution within the search space represents an animal position. The "fitness value" refers to the animal dominance with respect to the group. The complete process mimics the collective animal behavior.

The approach in this chapter implements a memory for storing best solutions (animal positions) mimicking the aforementioned biologic process. Such memory is divided into two different elements, one for maintaining the best locations at each generation ($\mathbf{M}_g$) and the other for storing the best historical positions during the complete evolutionary process ($\mathbf{M}_h$).

4.3.1 Description of the CAB Algorithm

Following other metaheuristic approaches, the CAB algorithm is an iterative process that starts by initializing the population randomly (generated random solutions or animal positions). Then, the following four operations are applied until a termination criterion is met (i.e. the iteration number NI):

1. Keep the position of the best individuals.
2. Move from or to nearby neighbors (local attraction and repulsion).
3. Move randomly.
4. Compete for the space within a determined distance (update the memory).

4.3.1.1 Initializing the Population

The algorithm begins by initializing a set A of N_p animal positions $\left(\mathbf{A} = \left\{\mathbf{a}_1, \mathbf{a}_2, \ldots, \mathbf{a}_{N_p}\right\}\right)$. Each animal position $\mathbf{a}_1$ is a D-dimensional vector containing parameter values to be optimized. Such values are randomly and uniformly distributed between the pre-specified lower initial parameter bound a_j^{low} and the upper initial parameter bound a_j^{high}.

$$a_{i,j} = a_j^{low} + \text{rand}(0,1) \cdot \left(a_j^{high} - a_j^{low}\right); \\ j = 1, 2, \ldots, D; \quad i = 1, 2, \ldots, N_p \tag{4.1}$$

with j and i being the parameter and individual indexes respectively. Hence, $a_{i,j}$ is the jth parameter of the ith individual.

All the initial positions A are sorted according to the fitness function (dominance) to form a new individual set $\mathbf{X} = \{\mathbf{x}_1, \mathbf{x}_2, \ldots, \mathbf{x}_{N_p}\}$, so that we can choose the best B positions and store them in the memory $\mathbf{M}_g$ and $\mathbf{M}_h$. The fact that both memories share the same information is only allowed at this initial stage.

4.3.1.2 Keep the Position of the Best Individuals

Analogous to the biological metaphor, this behavioral rule, typical from animal groups, is implemented as an evolutionary operation in our approach. In this operation, the first B elements $(\{\mathbf{a}_1, \mathbf{a}_2, \ldots, \mathbf{a}_B\})$, of the new animal position set A, are generated. Such positions are computed by the values contained inside the historical memory $\mathbf{M}_h$, considering a slight random perturbation around them. This operation can be modeled as follows:

$$\mathbf{a}_l = \mathbf{m}_h^l + \mathbf{v} \tag{4.2}$$

where $l \in \{1, 2, \ldots, B\}$ while $\mathbf{m}_h^l$ represents the l-element of the historical memory $\mathbf{M}_h$. $\mathbf{v}$ is a random vector with a small enough length.

4.3.1.3 Move from or to Nearby Neighbors

From the biological inspiration, animals experiment a random local attraction or repulsion according to an internal motivation. Therefore, we have implemented new evolutionary operators that mimic such biological pattern. For this operation, a uniform random number r_m is generated within the range [0, 1]. If r_m is less than a threshold H, a determined individual position is moved (attracted or repelled) considering the nearest best historical position within the group (i.e. the nearest position in $\mathbf{M}_h$); otherwise, it goes to the nearest best location within the group for the current generation (i.e. the nearest position in $\mathbf{M}_g$). Therefore such operation can be modeled as follows:

$$\mathbf{a}_i = \begin{cases} \mathbf{x}_i \pm r \cdot (\mathbf{m}_h^{nearest} - \mathbf{x}_i) & \text{with probability } H \\ \mathbf{x}_i \pm r \cdot (\mathbf{m}_g^{nearest} - \mathbf{x}_i) & \text{with probability } (1 - H) \end{cases} \tag{4.3}$$

where $i \in \{B+1, B+2, \ldots, N_p\}$, $\mathbf{m}_h^{nearest}$ and $\mathbf{m}_g^{nearest}$ represent the nearest elements of $\mathbf{M}_h$ and $\mathbf{M}_g$ to $\mathbf{x}_i$, while r is a random number.

4.3.1.4 Move Randomly

Following the biological model, under some probability P, one animal randomly changes its position. Such behavioral rule is implemented considering the next expression:

$$\mathbf{a}_i = \begin{cases} \mathbf{r} & \text{with probability } P \\ \mathbf{x}_i & \text{with probability } (1-P) \end{cases} \tag{4.4}$$

being $i \in \{B+1, B+2, \ldots, N_p\}$ and $\mathbf{r}$ a random vector defined in the search space. This operator is similar to re-initialize the particle in a random position, as it is done by Eq. (4.1).

4.3.1.5 Compete for the Space Within of a Determined Distance (Update the Memory)

Once the operations to keep the position of the best individuals, such as moving from or to nearby neighbors and moving randomly, have all been applied to the all N_p animal positions, generating N_p new positions, it is necessary to update the memory $\mathbf{M}_h$.

In order to update de memory $\mathbf{M}_h$, the concept of dominance is used. Animals that interact within a group maintain a minimum distance among them. Such distance ρ depends on how aggressive the animal behaves [34, 42]. Hence, when two animals confront each other inside such distance, the most dominant individual prevails meanwhile other withdraw. Figure 4.1 depicts the process.

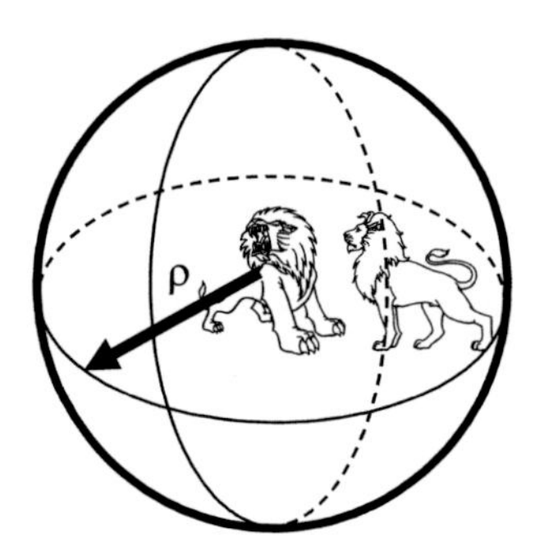

Fig. 4.1 Dominance concept as it is presented when two animals confront each other inside of a ρ distance

In the presented algorithm, the historical memory $\mathbf{M}_h$ is updated considering the following procedure:

1. The elements of $\mathbf{M}_h$ and $\mathbf{M}_g$ are merged into $\mathbf{M}_U$ $\left(\mathbf{M}_U = \mathbf{M}_h \cup \mathbf{M}_g\right)$.
2. Each element $\mathbf{m}_U^i$ of the memory $\mathbf{M}_U$ is compared pair-wise to remaining memory elements $\left(\left\{\mathbf{m}_U^1, \mathbf{m}_U^2, \ldots, \mathbf{m}_U^{2B-1}\right\}\right)$. If the distance between both elements is less than ρ, the element getting a better performance in the fitness function prevails meanwhile the other is removed.
3. From the resulting elements of $\mathbf{M}_U$ (from step 2), it is selected the B best value to build the new $\mathbf{M}_h$.

Unsuitable values of ρ yield a lower convergence rate, a longer computational time, a larger function evaluation number, the convergence to a local maximum or to an unreliable solution. The ρ value is computed considering the following equation:

$$\rho = \frac{\prod_{j=1}^{D} \left(a_j^{high} - a_j^{low}\right)}{10 \cdot D} \tag{4.5}$$

where a_j^{low} and a_j^{high} represent the pre-specified lower and upper bound of the j-parameter respectively, in an D-dimensional space.

4.3.1.6 Computational Procedure

The computational procedure for the presented algorithm can be summarized as follows:

Step 1	Set the parameters N_p, B, H, P and NI
Step 2	Generate randomly the position set $\mathbf{A} = \{\mathbf{a}_1, \mathbf{a}_2, \ldots, \mathbf{a}_{N_p}\}$ using Eq. 4.1
Step 3	Sort A according to the objective function (dominance) to build $\mathbf{X} = \{\mathbf{x}_1, \mathbf{x}_2, \ldots, \mathbf{x}_{N_p}\}$
Step 4	Choose the first B positions of $\mathbf{X}$ and store them into the memory $\mathbf{M}_g$
Step 5	Update $\mathbf{M}_h$ according to Sect. 4.3.1.5 (during the first iteration: $\mathbf{M}_h = \mathbf{M}_g$)
Step 6	Generate the first B positions of the new solution set $\mathbf{A} = \{\mathbf{a}_1, \mathbf{a}_2, \ldots, \mathbf{a}_B\}$. Such positions correspond to the elements of $\mathbf{M}_h$ making a slight random perturbation around them $\mathbf{a}_l = \mathbf{m}_h^l + \mathbf{v}$; being v a random vector of a small enough length
Step 7	Generate the rest of the A elements using the attraction, repulsion and random movements

(continued)

(continued)

	for $i=B+1: N_p$
	if ($r_1 < P$) then
	attraction and repulsion movement
	{ if ($r_2 < H$) then
	$\mathbf{a}_i = \mathbf{x}_i \pm r \cdot (\mathbf{m}_h^{nearest} - \mathbf{x}_i)$
	else if
	$\mathbf{a}_i = \mathbf{x}_i \pm r \cdot (\mathbf{m}_g^{nearest} - \mathbf{x}_i)$
	}
	else if
	random movement
	{
	$\mathbf{a}_i = \mathbf{r}$
	}
	end for where $r_1, r_2, r \in \text{rand}(0,1)$

Step 8	If *NI* is completed, the process is finished; otherwise go back to step 3. The best value in $\mathbf{M}_h$ represents the global solution for the optimization problem

4.4 Experimental Results

A comprehensive set of 23 functions, collected from Refs. [43–53], has been used to test the performance of the presented approach. Tables 4.1, 4.2 and 4.3 present the benchmark functions used in the experimental study. Such functions are classified into three different categories: Unimodal test functions (Table 4.1), multimodal test functions (Table 4.2) and multimodal test functions with fix dimensions

Table 4.1 Unimodal test functions

Test function	S	f_{opt}
$f_1(\mathbf{x}) = \sum_{i=1}^{n} x_i^2$	$[-100, 100]^n$	0
$f_2(\mathbf{x}) = \sum_{i=1}^{n} \lvert x_i \rvert + \prod_{i=1}^{n} \lvert x_i \rvert$	$[-10, 10]^n$	0
$f_3(\mathbf{x}) = \sum_{i=1}^{n} \left(\sum_{j=1}^{i} x_j \right)^2$	$[-100, 100]^n$	0
$f_4(\mathbf{x}) = \max_i \{ \lvert x_i \rvert, 1 \le i \le n \}$	$[-100, 100]^n$	0
$f_5(\mathbf{x}) = \sum_{i-1}^{n-1} \left[100 \left(x_{i+1} - x_i^2 \right)^2 + (x_i - 1)^2 \right]$	$[-30, 30]^n$	0
$f_6(\mathbf{x}) = \sum_{i=1}^{n} (x_i + 0.5)^2$	$[-100, 100]^n$	0
$f_7(\mathbf{x}) = \sum_{i=1}^{n} i x_i^4 + rand(0, 1)$	$[-1.28, 1.28]^n$	0

Table 4.2 Multimodal test functions

Test function	S	f_{opt}
$f_8(\mathbf{x}) = \sum_{i=1}^{n} -x_i \sin\left(\sqrt{\lvert x_i \rvert}\right)$	$[-500, 500]^n$	$-418.98 * n$
$f_9(\mathbf{x}) = \sum_{i=1}^{n} \left[x_i^2 - 10\cos(2\pi x_i) + 10\right]$	$[-5.12, 5.12]^n$	0
$f_{10}(\mathbf{x}) = -20\exp\left(-0.2\sqrt{\frac{1}{n}\sum_{i=1}^{n} x_i^2}\right) - \exp\left(\frac{1}{n}\sum_{i=1}^{n}\cos(2\pi x_i)\right) + 20$	$[-32, 32]^n$	0
$f_{11}(\mathbf{x}) = \frac{1}{4000}\sum_{i=1}^{n} x_i^2 - \prod_{i=1}^{n}\cos\left(\frac{x_i}{\sqrt{i}}\right) + 1$	$[-600, 600]^n$	0
$f_{12}(\mathbf{x}) = \frac{\pi}{n}\left\{10\sin(\pi y_1) + \sum_{i=1}^{n-1}(y_i - 1)^2\left[1 + 10\sin^2(\pi y_{i+1})\right] + (y_n - 1)^2\right\} + \sum_{i=1}^{n} u(x_i, 10, 100, 4)$ $y_i = 1 + \frac{x_i + 1}{4} \quad u(x_i, a, k, m) = \begin{cases} k(x_i - a)^m & x_i > a \\ 0 & -a < x_i < a \\ k(-x_i - a)^m & x_i < -a \end{cases}$	$[-50, 50]^n$	0
$f_{13}(\mathbf{x}) = 0.1\left\{\sin^2(3\pi x_1) + \sum_{i=1}^{n}(x_i - 1)^2\left[1 + \sin^2(3\pi x_i + 1)\right] + (x_n - 1)^2\left[1 + \sin^2(2\pi x_n)\right]\right\} + \sum_{i=1}^{n} u(x_i, 5, 100, 4)$	$[-50, 50]^n$	0

Table 4.3 Multimodal test functions with fix dimensions

Test function	S	f_{opt}
$f_{14}(\mathbf{x}) = \left(\frac{1}{500} + \sum_{j=1}^{25} \frac{1}{j + \sum_{i=1}^{2} (x_i - a_{ij})^6}\right)^{-1}$ $a_{ij} = \begin{pmatrix} -32, -16, 0, 16, 32, -32, \ldots, 0, 16, 32 \\ -32, -32, -32, -32, 16, \ldots, 32, 32, 32 \end{pmatrix}$	$[-65.5, 65.5]^2$	1
$f_{15}(\mathbf{x}) = \sum_{i=1}^{11} \left[a_i - \frac{x_i(b_i^2 + b_i x_2)}{b_i^2 + b_i x_3 + x_4}\right]^2$ $\mathbf{a} = [0.1957, 0.1947, 0.1735, 0.1600, 0.0844, 0.0627, 0.0456, 0.0342, 0.0342, 0.0235, 0.0246]$ $\mathbf{b} = [0.25, 0.5, 1, 2, 4, 6, 8, 10, 12, 14, 16]$	$[-5, 5]^4$	0.00030
$f_{16}(x_1, x_2) = 4x_1^2 - 2.1x_1^4 + \frac{1}{3}x_1^6 + x_1 x_2 - 4x_2^2 + 4x_2^4$	$[-5, 5]^2$	−1.0316
$f_{17}(x_1, x_2) = \left(x_2 - \frac{5.1}{4\pi^2}x_1^2 + \frac{5}{\pi}x_1 - 6\right)^2 + 10\left(1 - \frac{1}{8\pi}\right)\cos x_1 + 10$	$x_1 \in [-5, 10] x_2 \in [0, 15]$	0.398
$f_{18}(x_1, x_2) = \left[1 + (x_1 + x_2 + 1)^2(19 - 14x_1 + 3x_1^2 - 14x_2 + 6x_1x_2 + 3x_2^2\right]$ $\times \left[30 + (2x_1 - 3x_2)^2 \times (18 - 32x_1 + 12x_1^2 + 48x_2 - 36x_1x_2 + 27x_2^2\right]$	$[-5, 5]^2$	3
$f_{19}(\mathbf{x}) = -\sum_{i=1}^{4} c_i \exp\left(-\sum_{j=1}^{3} a_{ij}(x_j - p_{ij})^2\right)$ $\mathbf{a} = \begin{bmatrix} 3 & 10 & 30 \\ 0.1 & 10 & 35 \\ 3 & 10 & 30 \\ 0.1 & 10 & 30 \end{bmatrix}$ $\mathbf{c} = [1, 1.2, 3, 3.2]$ $\mathbf{P} = \begin{bmatrix} 0.3689 & 0.117 & 0.2673 \\ 0.4699 & 0.4387 & 0.7470 \\ 0.1091 & 0.8732 & 0.5547 \\ 0.0381 & 0.5743 & 0.8828 \end{bmatrix}$	$[0, 1]^3$	−3.86

(continued)

Table 4.3 (continued)

Test function	S	f_{opt}
$f_{20}(\mathbf{x}) = -\sum_{i=1}^{4} c_i \exp\left(-\sum_{j=1}^{6} a_{ij}(x_j - p_{ij})^2\right)$ $\mathbf{a} = \begin{bmatrix} 10 & 3 & 17 & 3.5 & 1.7 & 8 \\ 0.05 & 10 & 17 & 0.1 & 8 & 14 \\ 3 & 3.5 & 17 & 10 & 17 & 8 \\ 17 & 8 & 0.05 & 10 & 0.1 & 14 \end{bmatrix}$ $\mathbf{c} = [1, 1.2, 3, 3.2]$ $\mathbf{P} = \begin{bmatrix} 0.131 & 0.169 & 0.556 & 0.012 & 0.828 & 0.588 \\ 0.232 & 0.413 & 0.830 & 0.373 & 0.100 & 0.999 \\ 0.234 & 0.141 & 0.352 & 0.288 & 0.304 & 0.665 \\ 0.404 & 0.882 & 0.873 & 0.574 & 0.109 & 0.038 \end{bmatrix}$	$[0, 1]^6$	−3.32
$f_{21}(\mathbf{x}) = -\sum_{i=1}^{5} \left[(\mathbf{x} - a_i)(\mathbf{x} - a_i)^T + c_i\right]^{-1}$ $\mathbf{a} = \begin{bmatrix} 4 & 4 & 4 & 4 \\ 1 & 1 & 1 & 1 \\ 8 & 8 & 8 & 8 \\ 6 & 6 & 6 & 6 \\ 3 & 7 & 3 & 7 \\ 2 & 9 & 2 & 9 \\ 5 & 5 & 3 & 3 \\ 8 & 1 & 8 & 1 \\ 6 & 2 & 6 & 2 \\ 7 & 3.6 & 7 & 3.6 \end{bmatrix}$ $\mathbf{c} = [0.1, 0.2, 0.2, 0.4, 0.4, 0.6, 0.3, 0.7, 0.5, 0.5]$	$[0, 10]^4$	−10.1532
$f_{22}(\mathbf{x}) = -\sum_{i=1}^{7} \left[(\mathbf{x} - a_i)(\mathbf{x} - a_i)^T + c_i\right]^{-1}$ a and c, equal to f_{21}	$[0, 10]^4$	−10.4028
$f_{23}(\mathbf{x}) = -\sum_{i=1}^{7} \left[(\mathbf{x} - a_i)(\mathbf{x} - a_i)^T + c_i\right]^{-1}$ a and c, equal to f_{21}	$[0, 10]^4$	−10.5363

Table 4.4 Optimum locations of Table 4.3

Test function	$\mathbf{x}_{opt}$
f_{14}	(−32,32)
f_{15}	(0.1928,0.1908,0.1231,0.1358)
f_{16}	(0.089,−0.71), (−0.0089,0.712)
f_{17}	(−3.14,12.27), (3.14,2.275), (9.42,2.42)
f_{18}	(0,−1)
f_{19}	(0.114,0.556,0.852)
f_{20}	(0.201,0.15,0.477,0.275,0.311,0.657)
f_{21}	5 local minima in a_{ij}, $i = 1,\ldots,5$
f_{22}	7 local minima in a_{ij}, $i = 1,\ldots,7$
f_{23}	10 local minima in a_{ij}, $i = 1,\ldots,10$

(Table 4.2). In these tables, n is the dimension of function, f_{opt} is the minimum value of the function, and S is a subset of R^n. The optimum location ($\mathbf{x}_{opt}$) for functions in Tables 4.1 and 4.2, are in $[0]^n$, except for f_5, f_{12} and f_{13} with $\mathbf{x}_{opt}$ in $[1]^n$ and f_8 in $[420.96]^n$. A detailed description of optimum locations is given in Table 4.4.

4.4.1 Effect of the CAB Parameters

To study the impact of parameters P and H (described on Sects. 4.3.1.3 and 4.3.1.4) over the performance of CAB, different values have been tested on 5 typical functions. The maximum number of iterations is set to 1000. N_p and B are fixed to 50 and 10 respectively. The mean best function values (μ) and the standard deviations (σ^2) of CAB, averaged over 30 runs, for the different values of P and H are listed in Tables 4.5 and 4.6 respectively. The results suggest that a proper combination of different parameter values can improve the performance of CAB and the quality of solutions. Table 4.5 shows the results of an experiment which consist in fixing $H = 0.8$ and varying P from 0.5 to 0.9. On a second test, the experimental setup is swapped i.e. $P = 0.8$ and H varies from 0.5 to 0.9. The best results in the experiments are highlighted in Tables 4.5 and 4.6.

4.4.2 Performance Comparison

The CAB has been applied over the 23 functions comparing the results with those produced by Real Genetic Algorithm (RGA) [54], the PSO [18], the Gravitational Search Algorithm (GSA) [55] and the Differential Evolution method (DE) [56]. In all cases, population size is set to 50. The maximum iteration number is 1000 for functions in Tables 4.1 and 4.2 and 500 for functions in Table 4.3. Such stop criteria have been chosen as to keep compatibility to similar works in [14, 57].

Table 4.5 Results of CAB with variant values of parameter P over 5 typical functions, with $H = 0.8$

Function	n	$P = 0.5, \mu(\sigma^2)$	$P = 0.6, \mu(\sigma^2)$	$P = 0.7, \mu(\sigma^2)$	$P = 0.8, \mu(\sigma^2)$	$P = 0.9, \mu(\sigma^2)$
f_1	30	$2.63 \times 10^{-11}(2.13 \times 10^{-12})$	$1.98 \times 10^{-17}(6.51 \times 10^{-18})$	$1.28 \times 10^{-23}(3.54 \times 10^{-18})$	$\mathbf{2.33 \times 10^{-29}(4.41 \times 10^{-30})}$	$4.53 \times 10^{-23}(5.12 \times 10^{-24})$
f_3	30	$5.71 \times 10^{-13}(1.11 \times 10^{-14})$	$7.78 \times 10^{-19}(1.52 \times 10^{-20})$	$4.47 \times 10^{-27}(3.6 \times 10^{-28})$	$\mathbf{7.62 \times 10^{-31}(4.23 \times 10^{-32})}$	$3.42 \times 10^{-26}(3.54 \times 10^{-27})$
f_5	30	$5.68 \times 10^{-11}(2.21 \times 10^{-12})$	$1.54 \times 10^{-17}(1.68 \times 10^{-18})$	$5.11 \times 10^{-22}(4.42 \times 10^{-23})$	$\mathbf{9.02 \times 10^{-28}(6.77 \times 10^{-29})}$	$4.77 \times 10^{-20}(1.94 \times 10^{-21})$
f_{10}	30	$3.50 \times 10^{-5}(3.22 \times 10^{-6})$	$2.88 \times 10-^{9}(3.28 \times 10^{-10})$	$2.22 \times 10^{-12}(4.21 \times 10^{-13})$	$\mathbf{8.88 \times 10^{-16}(3.49 \times 10^{-17})}$	$1.68 \times 10^{-11}(5.31 \times 10^{-12})$
f_{11000}	30	$1.57 \times 10^{-12}(1.25 \times 10^{-3})$	$1.14 \times 10^{-6}(3.71 \times 10^{-7})$	$2.81 \times 10^{-8}(5.21 \times 10^{-9})$	$\mathbf{4.21 \times 10^{-10}(4.87 \times 10^{-11})}$	$4.58 \times 10^{-4}(6.92 \times 10^{-5})$

Table 4.6 Results of CAB with variant values of parameter H over 5 typical functions, with $P = 0.8$

Function	n	H = 0.5, $\mu(\sigma^2)$	H = 0.6, $\mu(\sigma^2)$	H = 0.7, $\mu(\sigma^2)$	H = 0.8, $\mu(\sigma^2)$	H = 0.9, $\mu(\sigma^2)$
f_1	30	$2.23 \times 10^{-10}(8.92 \times 10^{-11})$	$3.35 \times 10^{-18}(3.21 \times 10^{-19})$	$3.85 \times 10^{-22}(6.78 \times 10^{-23})$	$\mathbf{2.33 \times 10^{-29}(4.41 \times 10^{-30})}$	$4.72 \times 10^{-21}(6.29 \times 10^{-22})$
f_3	30	$5.71 \times 10^{-10}(5.12 \times 10^{-11})$	$3.24 \times 10^{-18}(1.32 \times 10^{-19})$	$6.29 \times 10^{-27}(8.26 \times 10^{-23})$	$\mathbf{7.62 \times 10^{-31}(4.23 \times 10^{-32})}$	$5.41 \times 10^{-22}(5.28 \times 10^{-23})$
f_5	30	$8.80 \times 10^{-9}(5.55 \times 10^{-10})$	$6.72 \times 10^{-21}(1.11 \times 10^{-22})$	$1.69 \times 10^{-23}(1.34 \times 10^{-24})$	$\mathbf{9.02 \times 10^{-28}(6.77 \times 10^{-29})}$	$7.39 \times 10^{-21}(4.41 \times 10^{-22})$
f_{10}	30	$2.88 \times 10^{-4}(3.11 \times 10^{-5})$	$3.22 \times 10^{-22}(2.18 \times 10^{-12})$	$1.23 \times 10^{-14}(4.65 \times 10^{-15})$	$\mathbf{8.88 \times 10^{-16}(3.49 \times 10^{-17})}$	$5.92 \times 10–^{7}(3.17 \times 10^{-9})$
f_{11}	30	$1.81 \times 10^{-4}(2.16 \times 10^{-5})$	$2.89 \times 10^{-6}(6.43 \times 10^{-7})$	$2.36 \times 10^{-7}(3.75 \times 10^{-4})$	$\mathbf{4.21 \times 10^{-10}(4.87 \times 10^{-11})}$	$3.02 \times 10^{-4}(4.37 \times 10^{-6})$

The parameter settings for each of the algorithms in the comparison are described as follows:

1. RGA: According to [54], the approach uses arithmetic crossover, Gaussian mutation and roulette wheel selection. The crossover and mutation probabilities have been set to 0.3 and 0.1 respectively.
2. PSO: In the algorithm, $c_1 = c_2 = 2$ while the inertia factor (ω) is decreasing linearly from 0.9 to 0.2.
3. In GSA, G_0 is set to 100 and α is set to 20; T is the total number of iterations (set to 1000 for functions f_1–f_{13} and to 500 for functions f_{14}–f_{23}). Besides, K_0 is set to 50 (total number of agents) and is decreased linearly to 1. Such values have been found as the best configuration set according to [55].
4. DE: The DE/Rand/1 scheme is employed. The parameter settings follow the instructions in [56]. The crossover probability is $CR = 0.9$ and the weighting factor is $F = 0.8$.

Unimodal Test Functions

Functions f_1 to f_7 are unimodal functions. The results for unimodal functions, over 30 runs, are reported in Table 4.7 considering the following performance indexes: the average best-so-far solution, the average mean fitness function and the median of the best solution in the last iteration. The best result for each function is boldfaced. According to this table, CAB provides better results than RGA, PSO, GSA and DE for all functions. In particular this test yields the largest difference in performance which is directly related to a better trade-off between exploration and exploitation produced by CAB operators. Moreover, the good convergence rate of CAB can be observed from Fig. 4.2. According to this figure, CAB tends to find the global optimum faster than other algorithms and yet offering the highest convergence rate.

A non-parametric statistical significance proof known as the Wilcoxon's rank sum test for independent samples [58, 59] has been conducted with an 5 % significance level, over the "average best-so-far" data of Table 4.7. Table 4.8 reports the p-values produced by Wilcoxon's test for the pair-wise comparison of the "average best so-far" of four groups. Such groups are formed by CAB versus RGA, CAB versus PSO, CAB versus GSA and CAB versus DE. As a null hypothesis, it is assumed that there is no significant difference between mean values of the two algorithms. The alternative hypothesis considers a significant difference between the "average best-so-far" values of both approaches. All p-values reported in the table are less than 0.05 (5 % significance level) which is a strong evidence against the null hypothesis, indicating that the CAB results are statistically significant and that it has not occurred by coincidence (i.e. due to the normal noise contained in the process).

Multimodal Test Functions

Multimodal functions have many local minima, being the most difficult to optimize. For multimodal functions, the final results are more important since they reflect the algorithm's ability to escape from poor local optima and locate a near-global

Table 4.7 Minimization result of benchmark functions in Table 4.1 with $n = 30$

		RGA	PSO	GSA	DE	CAB
f_1	Average best so-far	23.13	1.8×10^{-3}	7.3×10^{-11}	11.21	$\mathbf{2.3 \times 10^{-29}}$
	Median best so-far	21.87	1.2×10^{-3}	7.1×10^{-11}	13.21	$\mathbf{1.1 \times 10^{-20}}$
	Average mean fitness	23.45	1.2×10^{-2}	2.1×10^{-10}	11.78	$\mathbf{1.2 \times 10^{-20}}$
f_2	Average best so-far	1.07	2.0	4.03×10^{-5}	0.95	$\mathbf{5.28 \times 10^{-20}}$
	Median best so-far	1.13	1.9×10^{-3}	4.07×10^{-5}	1.05	$\mathbf{2.88 \times 10^{-11}}$
	Average mean fitness	1.07	2.0	6.9×10^{-5}	0.90	$\mathbf{1.43 \times 10^{-9}}$
f_3	Average best so-far	5.6×10^3	4.1×10^3	0.16×10^3	0.12	$\mathbf{7.62 \times 10^{-31}}$
	Median best so-far	5.6×10^3	2.2×10^3	0.15×10^3	0.09	$\mathbf{1.28 \times 10^{-19}}$
	Average mean fitness	5.6×10^3	2.9×10^3	0.16×10^3	0.11	$\mathbf{3.51 \times 10^{-12}}$
f_4	Average best so-far	11.78	8.1	3.7×10^{-6}	0.012	$\mathbf{2.17 \times 10^{-17}}$
	Median best so-far	11.94	7.4	3.7×10^{-6}	0.058	$\mathbf{5.65 \times 10^{-12}}$
	Average mean fitness	11.78	23.6	8.5×10^{-6}	0.013	$\mathbf{4.96 \times 10^{-10}}$
f_5	Average best so-far	1.1×10^3	3.6×10^4	25.16	0.25	$\mathbf{9.025 \times 10^{-28}}$
	Median best so-far	1.0×10^3	1.7×10^3	25.18	0.31	$\mathbf{3.10 \times 10^{-18}}$
	Average mean fitness	1.1×10^3	3.7×10^4	25.16	0.24	$\mathbf{6.04 \times 10^{-14}}$
f_6	Average best so-far	24.01	1.0×10^{-3}	8.3×10^{-11}	1.25×10^{-3}	$\mathbf{4.47 \times 10^{-29}}$
	Median best so-far	24.55	6.6×10^{-3}	7.7×10^{-11}	3.33×10^{-3}	$\mathbf{4.26 \times 10^{-21}}$
	Average mean fitness	24.52	0.02	2.6×10^{-10}	1.27×10^{-3}	$\mathbf{1.03 \times 10^{-12}}$
f_7	Average best so-far	0.06	0.04	0.018	6.87×10^{-3}	$\mathbf{3.45 \times 10^{-5}}$
	Median best so-far	0.06	0.04	0.015	4.72×10^{-3}	$\mathbf{7.39 \times 10^{-4}}$
	Average mean fitness	0.56	1.04	0.533	1.28×10^{-2}	$\mathbf{8.75 \times 10^{-4}}$

Maximum number of iterations = 1000

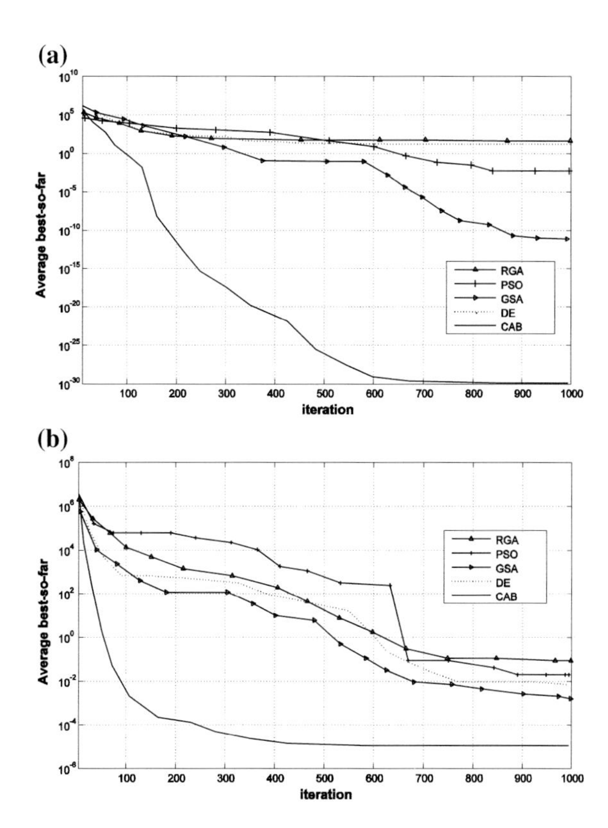

Fig. 4.2 Performance comparison of RGA, PSO, GSA, DE and CAB for minimization of **a** f_1 and **b** f_7 considering $n = 30$

optimum. Experiments have been done on f_8–f_{13} where the number of local minima increases exponentially as the dimension of the function increases. The dimension of these functions is set to 30. The results are averaged over 30 runs, reporting the performance indexes in Table 4.9 as follows: the average best-so-far solution, the average mean fitness function and the median of the best solution in the last iteration (the best result for each function is highlighted) Likewise, p-values of the Wilcoxon signed-rank test of 30 independent runs are listed in Table 4.10. For f_9, f_{10}, f_{11} and f_{12}, CAB yields a much better solution than the others. However, for functions f_8 and f_{13}, CAB produces similar results to RGA and GSA respectively. The Wilcoxon rank test results, presented in Table 4.10, show that CAB performed better than RGA, PSO, GSA and DE considering the four problems f_9–f_{11}, whereas, from a statistical viewpoint, there is not difference in results between CAB and RGA for f_8, and between CAB and GSA for f_{13}. The progress of the "average best-so-far" solution over 30 runs for functions f_{10} and f_{12} are shown in Fig. 4.3.

Multimodal Test Functions with Fix Dimensions

Table 4.11 shows the comparison between CAB, RGA, PSO, GSA and DE on multimodal benchmark functions with fix dimensions of Table 4.3. The results show that RGA, PSO and GSA have similar solutions and performances are nearly

Table 4.8 p-values produced by Wilcoxon's test comparing CAB versus RGA, PSO, GSA and DE over the "average best-so-far" values

CAB versus	RGA	PSO	GSA	DE
f_1	1.21×10^{-6}	3.94×10^{-5}	7.39×10^{-4}	1.04×10^{-6}
f_2	2.53×10^{-6}	5.62×10^{-5}	4.92×10^{-4}	2.21×10^{-6}
f_3	8.34×10^{-8}	6.42×10^{-8}	7.11×10^{-7}	1.02×10^{-4}
f_4	3.81×10^{-8}	1.91×10^{-8}	7.39×10^{-4}	1.27×10^{-6}
f_5	4.58×10^{-8}	9.77×10^{-9}	4.75×10^{-7}	0.23×10^{-4}
f_6	8.11×10^{-8}	1.98×10^{-6}	5.92×10^{-4}	2.88×10^{-5}
f_7	5.12×10^{-7}	4.77×10^{-7}	8.93×10^{-6}	1.01×10^{-4}

Table 4.9 Minimization result of benchmark functions in Table 4.2 with $n = 30$

		RGA	PSO	GSA	DE	CAB
f_8	Average best so-far	$\mathbf{-1.26 \times 10^4}$	-9.8×10^3	-2.8×10^3	-4.1×10^3	$\mathbf{-1.2 \times 10^4}$
	Median best so-far	$\mathbf{-1.26 \times 10^4}$	-9.8×10^3	-2.6×10^3	-4.1×10^3	$\mathbf{-1.2 \times 10^4}$
	Average mean fitness	$\mathbf{-1.26 \times 10^4}$	-9.8×10^3	-1.1×10^3	-4.1×10^3	$\mathbf{-1.2 \times 10^4}$
f_9	Average best so-far	5.90	55.1	15.32	30.12	$\mathbf{1.0 \times 10^{-3}}$
	Median best so-far	5.71	56.6	14.42	31.43	$\mathbf{7.6 \times 10^{-4}}$
	Average mean fitness	5.92	72.8	15.32	30.12	$\mathbf{1.0 \times 10^{-3}}$
f_{10}	Average best so-far	2.13	9.0×10^{-3}	6.9×10^{-6}	3.1×10^{-3}	$\mathbf{8.88 \times 10^{-16}}$
	Median best so-far	2.16	6.0×10^{-3}	6.9×10^{-6}	2.3×10^{-3}	$\mathbf{2.97 \times 10^{-11}}$
	Average mean fitness	2.15	0.02	1.1×10^{-5}	3.1×10^{-3}	$\mathbf{9.0 \times 10^{-10}}$
f_{11}	Average best so-far	1.16	0.01	0.29	1.0×10^{-3}	$\mathbf{1.14 \times 10^{-13}}$
	Median best so-far	1.14	0.0081	0.04	1.0×10^{-3}	$\mathbf{1.14 \times 10^{-13}}$
	Average mean fitness	1.16	0.055	0.29	1.0×10^{-3}	$\mathbf{1.14 \times 10^{-13}}$
f_{12}	Average best so-far	0.051	0.29	0.01	0.12	$\mathbf{2.32 \times 10^{-30}}$
	Median best so-far	0.039	0.11	4.2×10^{-13}	0.01	$\mathbf{5.22 \times 10^{-22}}$
	Average mean fitness	0.053	9.3×10^3	0.01	0.12	$\mathbf{4.63 \times 10^{-17}}$

(continued)

Table 4.9 (continued)

		RGA	PSO	GSA	DE	CAB
f_{13}	Average best so-far	0.081	3.1×10^{-18}	$\mathbf{3.2 \times 10^{-32}}$	1.77×10^{-25}	$\mathbf{1.35 \times 10^{-32}}$
	Median best so-far	0.032	2.2×10^{23}	$\mathbf{2.3 \times 10^{-32}}$	1.77×10^{-25}	$\mathbf{2.20 \times 10^{-21}}$
	Average mean fitness	0.081	4.8×10^{5}	$\mathbf{3.2 \times 10^{-32}}$	1.77×10^{-25}	$\mathbf{3.53 \times 10^{-17}}$

Maximum number of iterations = 1000

Table 4.10 p-values produced by Wilcoxon's test comparing CAB versus RGA, PSO, GSA and DE over the "average best-so-far" values from Table 4.9

CAB versus	RGA	PSO	GSA	DE
f_8	0.89	8.38×10^{-4}	1.21×10^{-4}	4.61×10^{-4}
f_9	7.23×10^{-7}	1.92×10^{-9}	5.29×10^{-8}	9.97×10^{-8}
f_{10}	6.21×10^{-9}	4.21×10^{-5}	1.02×10^{-4}	3.34×10^{-4}
f_{11}	7.74×10^{-9}	3.68×10^{-7}	4.10×10^{-7}	8.12×10^{-5}
f_{12}	1.12×10^{-8}	8.80×10^{-9}	2.93×10^{-7}	4.02×10^{-8}
f_{13}	4.72×10^{-9}	3.92×10^{-5}	0.93	2.20×10^{-4}

the same as it can be seen in Fig. 4.4. The results, presented in Table 4.11, show how metaheuristic algorithms maintain a similar average performance when they faced low-dimensional functions [57].

Comparison to Continuous Optimization Methods

Finally, CAB has been further compared to continuous optimization methods by considering the presented benchmark functions. Since the BFSG algorithm [60] is one of the most effective continuous methods for solving unconstrained optimization problems, it has been considered as a basis for the algorithms in the comparison. All experiments have been tested in MatLAB© over the same Dell Optiplex GX260 computer with a Pentium-4 2.66G-HZ processor, running Windows XP operating system over 1 Gb of memory.

Local Optimization

In the first experiment, the performance of algorithms BFGS and CAB over unimodal functions is compared. In unimodal functions, the global minimum matches the local minimum. Quasi-Newton methods, such as the BFGS, have a fast rate of local convergence despite it depends on the problem's dimension [61, 62]. Considering that not all unimodal functions of Table 4.1 fulfill the requirements imposed by the gradient based approaches (i.e. f_2 and f_4 are not differentiable meanwhile f_7 is non-smooth), we have chosen the Rosenbrock function (f_5) as a benchmark.

In the test, both algorithms (BFGS and CAB) are employed to minimize f_5, considering different dimensions. For the BFGS implementation, it is considered

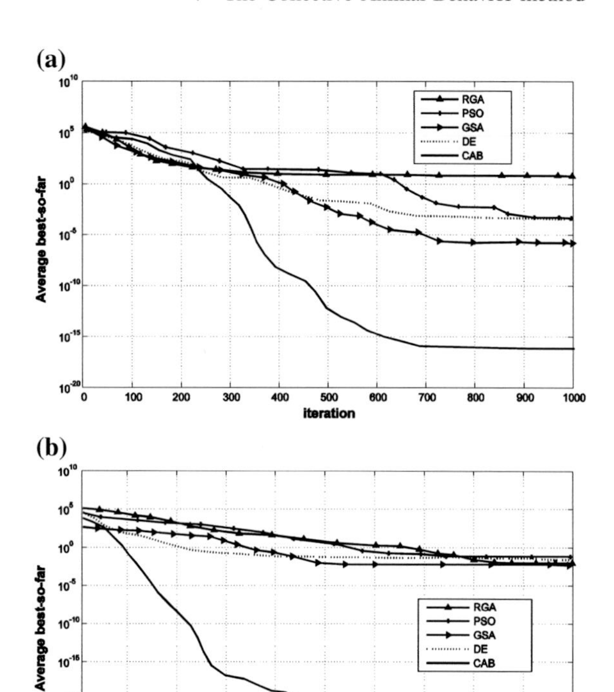

Fig. 4.3 Performance comparison of RGA, PSO, GSA, DE and CAB for minimization of **a** f_{10} and **b** f_{12} considering $n = 30$

Table 4.11 Minimization result of benchmark functions in Table 4.3 with $n = 30$

		RGA	PSO	GSA	DE	CAB
f_{14}	Average best so-far	0.998	0.998	3.70	0.998	0.998
	Median best so-far	0.998	0.998	2.07	0.998	0.998
	Average mean fitness	0.998	0.998	9.17	0.998	0.998
f_{15}	Average best so-far	4.0×10^{-3}	2.8×10^{-3}	8.0×10^{-3}	2.2×10^{-3}	1.1×10^{-3}
	Median best so-far	1.7×10^{-3}	7.1×10^{-4}	7.4×10^{-4}	5.3×10^{-4}	2.2×10^{-4}
	Average mean fitness	4.0×10^{-3}	215.60	9.0×10^{-3}	2.2×10^{-3}	1.1×10^{-3}
f_{16}	Average best so-far	−1.0313	−1.0316	−1.0316	−1.0316	−1.0316
	Median best so-far	−1.0315	−1.0316	−1.0316	−1.0316	−1.0316
	Average mean fitness	−1.0313	−1.0316	−1.0316	−1.0316	−1.0316

(continued)

Table 4.11 (continued)

		RGA	PSO	GSA	DE	CAB
f_{17}	Average best so-far	0.3996	0.3979	0.3979	0.3979	0.3979
	Median best so-far	0.3980	0.3979	0.3979	0.3979	0.3979
	Average mean fitness	0.3996	2.4112	0.3979	0.3979	0.3979
f_{18}	Average best so-far	−3.8627	−3.8628	−3.8628	−3.8628	−3.8628
	Median best so-far	−3.8628	−3.8628	−3.8628	−3.8628	−3.8628
	Average mean fitness	−3.8627	−3.8628	−3.8628	−3.8628	−3.8628
f_{19}	Average best so-far	−3.3099	−3.3269	−3.7357	−3.3269	−3.8501
	Median best so-far	−3.3217	−3.2031	−3.8626	−3.3269	−3.8501
	Average mean fitness	−3.3098	−3.2369	−3.8020	−3.3269	−3.8501
f_{20}	Average best so-far	−3.3099	−3.2369	−2.0569	−3.2369	−3.2369
	Median best so-far	−3.3217	−3.2031	−1.9946	−3.2369	−3.2369
	Average mean fitness	−3.3098	−3.2369	−1.6014	−3.2369	−3.2369
f_{21}	Average best so-far	−5.6605	−6.6290	−6.0748	−6.6290	−10.1532
	Median best so-far	−2.6824	−5.1008	−5.0552	−6.0748	−10.1532
	Average mean fitness	−5.6605	−5.7496	−6.0748	−6.6290	−10.1532
f_{22}	Average best so-far	−7.3421	−9.1118	−9.3339	−9.3339	−10.4028
	Median best so-far	−10.3932	−10.402	−10.402	−9.3339	−10.4028
	Average mean fitness	−7.3421	−9.9305	−9.3399	−9.3339	−10.4028
f_{23}	Average best so-far	−6.2541	−9.7634	−9.4548	−9.7634	−10.5363
	Median best so-far	−4.5054	−10.536	−10.536	−9.7636	−10.5363
	Average mean fitness	−6.2541	−8.7626	−9.4548	−9.7634	−10.5363

Maximum number of *iterations* = 500

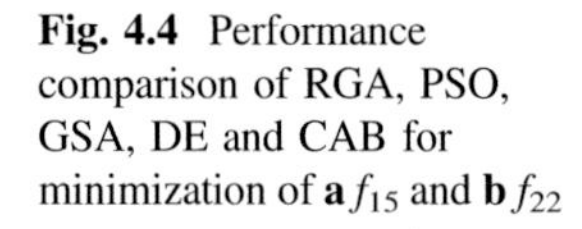

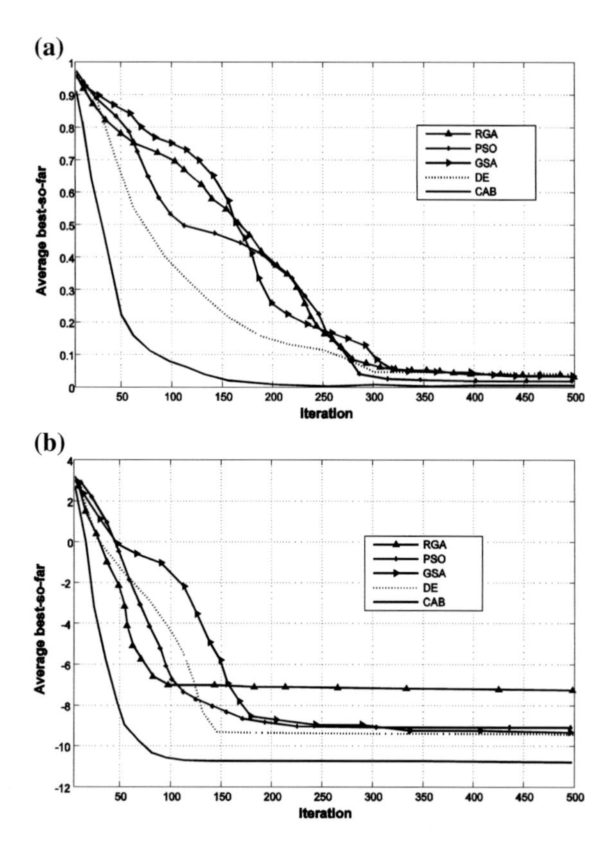

Fig. 4.4 Performance comparison of RGA, PSO, GSA, DE and CAB for minimization of **a** f_{15} and **b** f_{22}

$B_0 = I$ as initial matrix. Likewise, parameters δ and δ are set to 0.1 and 0.9 respectively. Although several performance criteria may define a comparison index, most can be applied to only one method timely (such as the number of gradient evaluations). Therefore, this chapter considers the elapsed time and the iteration number (once the minimum has been reached) as performance indexes in the comparison. In the case of BFGS, the termination condition is assumed as $\|g_5(\mathbf{x})\| \leq 1 \times 10^{-6}$, with $g_5(\mathbf{x})$ being the gradient of $g_5(\mathbf{x})$. On the other hand, the stopping criterion of CAB considers when no more changes to the best element in memory $\mathbf{M}_h$ are registered. Table 4.12 presents the results of both algorithms considering several dimensions ($n \in \{2, 10, 30, 50, 70, 100, 120\}$) of f_5. In order to assure consistency, such results represent the averaged elapsed time (AET) and the averaged iteration number (AIN) over 30 different executions. It is additionally considered that at each execution both methods are initialized to a random point (inside the search space).

From Table 4.12, we can observe that the BFGS algorithm produces shorter elapsed times and fewer iterations than the CAB method. However, from n = 70, the CAB algorithm contend with similar results. The fact that the BFGS algorithm

Table 4.12 Performance comparison between the BFGS and the CAB algorithm, considering different dimensions over the Rosenbrock function

f_5	AET		AIN	
n	BFGS	CAB	BFGS	CAB
2	0.15	4.21	6	89
10	0.55	5.28	22	98
30	1.35	5.44	41	108
50	2.77	5.88	68	112
70	4.23	6.11	93	115
100	5.55	6.22	105	121
120	6.64	6.71	125	129

The averaged elapsed time (AET) is referred in seconds

outperforms the CAB approach cannot be deemed as a negative feature considering the restrictions imposed to the functions by the BFGS method.

Global Optimization
Since the BFGS algorithm exploits only local information, it may easily get trap into local optima restricting its use for global optimization. Therefore several other methods for continuous optimization have been proposed. One of the most widely used techniques is the so called Multi-Start [63] (MS). In MS a point is randomly chosen from a feasible region as initial solution and subsequently a continuous optimization algorithm (local search) starts from it. Then, the process is repeated until a near global optimum is reached. The weakness of MS is that the same local minima may be found over and over again, wasting computational resources [64].

In order to compare the performance of the CAB approach to continuous optimization methods in the context of global optimization, the MS algorithm ADAPT [65] has been chosen. ADAPT uses an iterative BFGS algorithm as local search method. Thus, ADAPT possess two different stop criteria, one for the local procedure BFGS and other for the complete MS approach. For the comparison, the ADAPT algorithm has been implemented as suggested by [65].

In the second experiment, the performance of the ADAPT and the CAB algorithms is compared over several multimodal functions described in Tables 4.2 and 4.3. The study considers the following performance indexes: the elapsed time, the iteration number and the average best so-far solution. In case of the ADAPT algorithm, the iteration number is computed as the total iteration number produced by all the local search procedures as the MS method operates. The termination condition of the ADAPT local search algorithm (BFGS) is assumed when $\|g_k(\mathbf{x})\| \leq 1 \times 10^{-5}$, being $g_k(\mathbf{x})$ the gradient of $f_k(\mathbf{x})$. On the other hand, the stopping criterion for the CAB and the ADAPT algorithms, is considered when no more changes in the best element are registered, i.e. the best element in $\mathbf{M}_h$ for CAB or the best found-so-far element during the MS process. Table 4.13 presents results from both algorithms considering several multimodal functions. In order to assure consistency, results ponder the averaged elapsed time (AET), the averaged iteration number (AIN) and the average best so-far solution (ABS) over 30 different

Table 4.13 Performance comparison between the ADAPT and the CAB algorithm considering different multimodal functions

Function	n	ADAPT				CAB		
		ALS	AET	AIN	ABS	AET	AIN	ABS
f_9	30	3705	45.4	23,327	1.2×10^{-2}	10.2	633	1.0×10^{-3}
f_{10}	30	4054	1′05.7	38,341	6.21×10^{-12}	12.1	723	8.88×10^{-16}
f_{11}	30	32,452	2′12.1	102,321	4.51×10^{-10}	15.8	884	1.14×10^{-13}
f_{17}	2	1532	33.2	20,202	0.3976	7.3	332	0.3979
f_{18}	2	1233	31.6	18,845	−3.8611	6.6	295	−3.8628

The averaged elapsed time (AET) is referred in the format M's (Minute' second)

executions. In Table 4.13, the averaged number of local searches (ALS) executed by ADAPT during the optimization, is additionally considered.

Table 4.13 provides a summarized performance comparison between the ADAPT and the CAB algorithms. Despite both algorithms are able to acceptably locate the global minimum for both cases, there exist significant differences in the required time for reaching it. When comparing the averaged elapsed time (AET) and the averaged iteration number (AIN) in Table 4.13, CAB uses significantly less time and fewer iterations to reach the global minimum than the ADAPT algorithm.

4.5 Summary

In recent years, several metaheuristic optimization methods have been developed including some that have been inspired by nature-related phenomena. This chapter presents a novel optimization algorithm that is called the Collective Animal Behavior Algorithm (CAB). In CAB, the searcher agents emulate a group of animals that interact to each other based on simple behavioral rules which are modeled as mathematical operators. Such operations are applied to each agent considering that the complete group has a memory storing their own best positions seen so far by using a competition principle.

CAB has been experimentally tested considering a challenging test suite consisting of 23 benchmark functions. The performance of CAB has been compared against the following metaheuristic algorithms: the Real Genetic Algorithm (RGA) [54], the PSO [18], the Gravitational Search Algorithm (GSA) [55] and the Differential Evolution (DE) method [56]. The experiments have demonstrated that CAB generally outperforms other metaheuristic algorithms for most of the benchmark functions regarding the solution quality. In this study the CAB algorithm has also been compared to algorithms based in continuous optimization methods such as the BFGS [60] and the ADAPT [65]. The results shown that although BFGS outperforms CAB for local optimization tasks, ADAPT faces great difficulties in solving global optimization problems. CAB's remarkable performance is due to two different reasons: (i) the defined operators allow a better exploration of the search space, increasing the capacity to find multiple optima; and (ii) the diversity of the solutions contained in the $\mathbf{M}_h$ memory, in terms of multimodal optimization, is maintained and even improved by implementing the competition principle (dominance concept).

References

1. Pardalos Panos, M., Romeijn Edwin, H., Toy, H.: Recent developments and trends in global optimization. J. Comput. Appl. Math. **124**, 209–228 (2000)
2. Floudas, C., Akrotirianakis, I., Caratzoulas, S., Meyer, C., Kallrath, J.: Global optimization in the 21st century: advances and challenges. Comput. Chem. Eng. **29**(6), 1185–1202 (2005)
3. Ying, J., Ke-Cun, Z., Shao-Jian, Q.: A deterministic global optimization algorithm. Appl. Math. Comput. **185**(1), 382–387 (2007)
4. Georgieva, A., Jordanov, I.: Global optimization based on novel heuristics, low-discrepancy sequences and genetic algorithms. Eur. J. Oper. Res. **196**, 413–422 (2009)
5. Lera, D., Sergeyev, Y.: Lipschitz and Hölder global optimization using space-filling curves. Appl. Numer. Math. **60**(1–2), 115–129 (2010)
6. Fogel, L.J., Owens, A.J., Walsh, M.J.: Artificial Intelligence Through Simulated Evolution. Wiley, Chichester, UK (1966)
7. De Jong, K.: Analysis of the behavior of a class of genetic adaptive systems. Ph.D. thesis, University of Michigan, Ann Arbor, MI (1975)
8. Koza, J.R.: Genetic programming: a paradigm for genetically breeding populations of computer programs to solve problems. Rep. No. STAN-CS-90-1314, Stanford University, CA (1990)
9. Holland, J.H.: Adaptation in Natural and Artificial Systems. University of Michigan Press, Ann Arbor (1975)
10. Goldberg, D.E.: Genetic Algorithms in Search, Optimization and Machine Learning. Addison Wesley, Boston (1989)
11. de Castro L.N., Von Zuben F.J.: Artificial immune systems: Part I—basic theory and applications. Technical report, TR-DCA 01/99. December 1999
12. Kirkpatrick, S., Gelatt, C., Vecchi, M.: Optimization by simulated annealing. Science **220** (4598), 671–680 (1983)
13. İlker, B., Birbil, S., Shu-Cherng, F.: An electromagnetism-like mechanism for global optimization. J. Global Optim. **25**, 263–282 (2003)
14. Rashedia, E., Nezamabadi-pour, H., Saryazdi, S.: Filter modeling using gravitational search algorithm. Eng. Appl. Artif. Intell. **24**(1), 117–122 (2011)
15. Geem, Z.W., Kim, J.H., Loganathan, G.V.: A new heuristic optimization algorithm: harmony search. Simulation **76**(2), 60–68 (2001)
16. Lee, K.S., Geem, Z.W.: A new meta-heuristic algorithm for continues engineering optimization: harmony search theory and practice. Comput. Methods Appl. Mech. Eng. **194**, 3902–3933 (2004)
17. Geem, Z.W.: Novel derivative of harmony search algorithm for discrete design variables. Appl. Math. Comput. **199**, 223–230 (2008)
18. Kennedy, J., Eberhart, R.C.: Particle swarm optimization. In: Proceedings of the 1995 IEEE International Conference on Neural Networks, vol. 4, pp. 1942–1948 (1995)
19. Dorigo, M., Maniezzo, V., Colorni, A.: Positive feedback as a search strategy. Technical Report No. 91-016, Politecnico di Milano (1991)

20. Oftadeh, R., Mahjoob, M.J., Shariatpanahi, M.: A novel meta-heuristic optimization algorithm inspired by group hunting of animals: hunting search. Comput. Math Appl. **60**, 2087–2098 (2010)
21. Sumper, D.: The principles of collective animal behaviour. Philos. Trans. R. Soc. Lond. B Biol. Sci. **361**(1465), 5–22 (2006)
22. Petit, O., Bon, R.: Decision-making processes: the case of collective movements. Behav. Process. **84**, 635–647 (2010)
23. Kolpas, A., Moehlis, J., Frewen, T., Kevrekidis, I.: Coarse analysis of collective motion with different communication mechanisms. Math. Biosci. **214**, 49–57 (2008)
24. Couzin, I.: Collective cognition in animal groups. Trends Cogn. Sci. **13**(1), 36–43 (2008)
25. Couzin, I.D., Krause, J.: Self-organization and collective behavior in vertebrates. Adv. Stud. Behav. **32**, 1–75 (2003)
26. Bode, N., Franks, D., Wood, A.: Making noise: emergent stochasticity in collective motion. J. Theor. Biol. **267**, 292–299 (2010)
27. Couzi, I., Krause, I., James, R., Ruxton, G., Franks, N.: Collective memory and spatial sorting in animal groups. J. Theor. Biol. **218**, 1–11 (2002)
28. Couzin, I.D.: Collective minds. Nature **445**, 715–728 (2007)
29. Bazazi, S., Buhl, J., Hale, J.J., Anstey, M.L., Sword, G.A., Simpson, S.J., Couzin, I.D.: Collective motion and cannibalism in locust migratory bands. Curr. Biol. **18**, 735–739 (2008)
30. Bode, N., Wood, A., Franks, D.: The impact of social networks on animal collective motion. Anim. Behav. **82**(1), 29–38 (2011)
31. Lemasson, B., Anderson, J., Goodwin, R.: Collective motion in animal groups from a neurobiological perspective: the adaptive benefits of dynamic sensory loads and selective attention. J. Theor. Biol. **261**(4), 501–510 (2009)
32. Bourjade, M., Thierry, B., Maumy, M., Petit, O.: Decision-making processes in the collective movements of Przewalski horses families Equus ferus Przewalskii: influences of the environment. Ethology **115**, 321–330 (2009)
33. Banga, A., Deshpande, S., Sumanab, A., Gadagkar, R.: Choosing an appropriate index to construct dominance hierarchies in animal societies: a comparison of three indices. Anim. Behav. **79**(3), 631–636 (2010)
34. Hsu, Y., Earley, R., Wolf, L.: Modulation of aggressive behaviour by fighting experience: mechanisms and contest outcomes. Biol. Rev. **81**(1), 33–74 (2006)
35. Broom, M., Koenig, A., Borries, C.: Variation in dominance hierarchies among group-living animals: modeling stability and the likelihood of coalitions. Behav. Ecol. **20**, 844–855 (2009)
36. Bayly, K.L., Evans, C.S., Taylor, A.: Measuring social structure: a comparison of eight dominance indices. Behav. Process. **73**, 1–12 (2006)
37. Conradt, L., Roper, T.J.: Consensus decision-making in animals. Trends Ecol. Evol. **20**, 449–456 (2005)
38. Okubo, A.: Dynamical aspects of animal grouping. Adv. Biophys. **22**, 1–94 (1986)
39. Reynolds, C.W.: Flocks, herds and schools: a distributed behavioural model. Comp. Graph. **21**, 25–33 (1987)
40. Gueron, S., Levin, S.A., Rubenstein, D.I.: The dynamics of mammalian herds: from individual to aggregations. J. Theor. Biol. **182**, 85–98 (1996)
41. Czirok, A., Vicsek, T.: Collective behavior of interacting self-propelled particles. Phys. A **281**, 17–29 (2000)
42. Ballerini, M.: Interaction ruling collective animal behavior depends on topological rather than metric distance: evidence from a field study. Proc. Natl. Acad. Sci. USA **105**, 1232–1237 (2008)
43. Ali, M.M., Khompatraporn, C., Zabinsky, Z.B.: A numerical evaluation of several stochastic algorithms on selected continuous global optimization test problems. J. Global Optim. **31**(4), 635–672 (2005)
44. Chelouah, R., Siarry, P.: A continuous genetic algorithm designed for the global optimization of multimodal functions. J. Heuristics **6**(2), 191–213 (2000)

45. Herrera, F., Lozano, M., Sánchez, A.M.: A taxonomy for the crossover operator for real-coded genetic algorithms: an experimental study. Int. J. Intell. Syst. **18**(3), 309–338 (2003)
46. Laguna, M., Martí, R.: Experimental testing of advanced scatter search designs for global optimization of multimodal functions. J. Global Optim. **33**(2), 235–255 (2005)
47. Lozano, M., Herrera, F., Krasnogor, N., Molina, D.: Real-coded memetic algorithms with crossover hill-climbing. Evol. Comput. **12**(3), 273–302 (2004)
48. Moré, J.J., Garbow, B.S., Hillstrom, K.E.: Testing unconstrained optimization software. ACM Trans. Math. Softw. **7**(1), 17–41 (1981)
49. Ortiz-Boyer, D., Hervás-Martınez, C., García-Pedrajas, N.: CIXL2: a crossover operator for evolutionary algorithms based on population features. J. Artif. Intell. Res. **24**(1), 1–48 (2005)
50. Price, K., Storn, R.M., Lampinen, J.A.: Differential Evolution: A Practical Approach to Global Optimization. Springer, New York (2005)
51. Rahnamayan, S., Tizhoosh, H.R., Salama, M.M.A.: Opposition-based differential evolution. IEEE Trans. Evol. Comput. **12**(1), 64–79 (2008)
52. Whitley, D., Rana, D., Dzubera, J., Mathias, E.: Evaluating evolutionary algorithms. Artif. Intell. **85**(1–2), 245–276 (1996)
53. Yao, X., Liu, Y., Lin, G.: Evolutionary programming made faster. IEEE Trans. Evol. Comput. **3**(2), 82–102 (1999)
54. Hamzaçebi, C.: Improving genetic algorithms' performance by local search for continuous function optimization. Appl. Math. Comput. **196**(1), 309–317 (2008)
55. Rashedi, E., Nezamabadi-pour, H., Saryazdi, S.: GSA: a Gravitational Search Algorithm. Inf. Sci. **179**, 2232–2248 (2009)
56. Storn, R., Price, K.: Differential evolution—a simple and efficient adaptive scheme for global optimisation over continuous spaces. Technical Report TR-95–012, ICSI, Berkeley, CA (1995)
57. Shilane, D., Martikainen, J., Dudoit, S., Ovaska, S.: A general framework for statistical performance comparison of evolutionary computation algorithms. Inf. Sci. **178**, 2870–2879 (2008)
58. Wilcoxon, F.: Individual comparisons by ranking methods. Biometrics **1**, 80–83 (1945)
59. Garcia, S., Molina, D., Lozano, M., Herrera, F.: A study on the use of non-parametric tests for analyzing the evolutionary algorithms' behaviour: a casc study on the CEC'2005 Special session on real parameter optimization. J Heurist (2008). doi:10.1007/s10732-008-9080-4
60. Al-Baali, M.: On the behavior of bombined extra-updating/self scaling BFGS method. J. Comput. Appl. Math. **134**, 269–281 (2001)
61. Powell, M.: How bad are the BFGS and DFP methods when the objective function is quadratic? Math. Program. **34**, 34–37 (1986)
62. Hansen, E., Walster, G.: Global Optimization Using Interval Analysis. CRC Press, Boca Raton (2004)
63. Lasdona, L., Plummer, J.: Multistart algorithms for seeking feasibility. Comput. Oper. Res. **35** (5), 1379–1393 (2008)
64. Theos, F., Lagaris, I., Papageorgiou, D.: PANMIN: sequential and parallel global optimization procedures with a variety of options for the local search strategy. Comput. Phys. Commun. **159**, 63–69 (2004)
65. Voglis, C., Lagaris, I.: Towards "Ideal Multistart". A stochastic approach for locating the minima of a continuous function inside a bounded domain. Appl. Math. Comput. **213**, 216–229 (2009)

Chapter 5
An Evolutionary Computation Algorithm based on the Allostatic Optimization

5.1 Introduction

Global Optimization (GO) is a field with applications in many areas of science, engineering, economics and others, where mathematical modelling is used [1]. In general, the goal is to find a global optimum of an objective function defined in a given search space. Global optimization algorithms are usually broadly divided into deterministic and evolutionary [2]. Since deterministic methods only provide a theoretical guarantee of locating a local minimum of the objective function, they often face great difficulties in solving global optimization problems [3, 4]. On the other hand, evolutionary methods are usually faster in locating a global optimum than deterministic ones [5]. Moreover, evolutionary algorithms adapt better to black-box formulations and extremely ill-behaved functions, whereas deterministic methods usually rest on at least some theoretical assumptions about the problem formulation and its analytical properties (such as Lipschitz continuity) [3, 4].

Several evolutionary algorithms have been developed by a combination of deterministic rules and randomness mimicking several phenomena. Such phenomena include animal behavior e.g. the Particle Swarm Optimization (PSO) algorithm proposed by Kennedy and Eberhart [6], the Artificial Bee Colony (ABC) proposed by Karaboga [7] and the Ant Colony Optimization (ACO) algorithm proposed by Dorigo et al. [8]; physical processes such as the Simulated Annealing (SA) proposed by Kirkpatrick et al. [9], the Electromagnetism-like Algorithm proposed by İlker et al. [10], the Gravitational Search Algorithm (GSA) proposed by Rashedi et al. [11] and musical process such as the Harmony Search (HS) algorithm, which has been proposed by Geem et al. [12].

An impressive growth in the field of biologically inspired meta-heuristics for search and optimization has emerged during the last decade. Some bio-inspired examples include processes such as the Evolutionary Algorithm (EA) proposed by

E. Cuevas et al., *Advances of Evolutionary Computation: Methods and Operators*, Studies in Computational Intelligence 629,
DOI 10.1007/978-3-319-28503-0_5

Fogel et al. [13], De Jong [14], and Koza [15], the Genetic Algorithms (GA) proposed by Holland [16], the Artificial Immune System proposed by de Castro and Von Zuben [17] and the Differential Evolution Algorithm (DE) proposed by Storn and Price [18].

Although DE, ABC and PSO algorithms have become popular in solving complex optimization problem, they present serious flaws such as premature convergence and difficulty to overcome local minima. The reason of these problems is the operators used for modifying the particles. In such algorithms, during their evolution, the position of each agent in the next iteration is updated yielding an attraction towards the position of the best particle seen so-far [19, 20]. This behavior produces that the entire population, as the algorithm evolves, concentrates around the best particle, favoring the premature convergence and damaging the particle diversity.

Allostasis is a biological term which explains how the modifications of specialized organ conditions inside the body allow to achieve stability when an unbalance health condition is presented. If a body decompensation happens, according to the allostatic mechanisms, several body conditions compound by blood pressure, oxygen tension and others indexes are proved in order to get a stability state. Such body conditions are generated by using different specialized mechanisms.

In this chapter, a novel biologically-inspired algorithm, namely Allostatic Optimization (AO) is presented for solving optimization tasks. The AO algorithm uses as metaphor the allostasis mechanisms for designing an optimization methodology. AO provides a population-based search procedure under which all individuals, also seen as body conditions, are defined within the multidimensional space. They are generated or modified by using several evolutionary operators that emulate the different operations employed by the allostasis process whereas the fitness function evaluates the capacity of each individual (body condition) to reach a steady health state (good solution). Different to DE, ABC and PSO, the presented algorithm implements evolutionary operators that avoid concentrating most of the particles in only one position favoring the exploration process and eliminating the flaws related to the premature convergence. Our approach has been compared to other well-known evolutionary algorithms. The results confirm a high performance of the presented method for solving various benchmark functions.

The rest of this chapter is organized as follows: in Sect. 5.2 it is presented a review of the biological source of inspiration, allostasis, as well as its associated mechanisms, whereas in Sect. 5.3 it is shown in detail the presented algorithm; Sect. 5.4 is devoted to the experimental part, where the performance of the presented approach is evaluated using several benchmark functions, and in Sect. 5.5 some conclusions related to the proposal are drawn.

5.2 Allostasis, a Brief Review

Researchers had studied for a century several human diseases, always considering that each organ could independently reach a recovery state, without information from other organs [21–23]. In such a way, if a person had a problem with a system as, say, heart, then thinking that only a single body condition (such as blood pressure) had the whole responsibility of returning the system to stability, it is prescribed medicine for increasing or decreasing the blood pressure. Such way of explaining the recovery process, it is call homeostasis.

On the other hand, allostasis is a recent biological term which explains how the modifications of different organ conditions inside the body allow to achieve stability when an unbalance health condition is presented. Different to homeostasis, allostasis considers the interaction of several body conditions (instead of only one) in order to explain the recovery body process. Under the allostasis mechanism, if a body decompensation happens, several body conditions compound by blood pressure, oxygen tension, glucose level, hormones and others indexes are proved in order to get a stability state. Such body conditions are generated by using different specialized mechanisms. Figure 5.1 shows graphically the allostasis mechanism.

In alloestasis, a complex mechanism is used in order to generate new body conditions. Such process considers a previous set of body conditions which have been historically used. The set is divided in two different groups: Group A and Group B. Body conditions of Group A do not suffer collective changes whereas those contained in Group B are constantly modified by collective operations.

Several procedures are considered by allostasis in order to generate new body conditions. The main mechanism is called "combination". It combines each body condition with information provided by other organs. In this mechanism, a body condition, taken from the set of body conditions, is combined with information of other organs to produce a new body condition. Once a new body condition is generated, it is evaluated its capacity to reach a steady health state. If the new body condition improves the health sate provided by the last body condition, a collective change is carried out over all elements contained in Group B. Such collective changes are selectively applied to Group B considering three different sub-sets namely Group B1, Group B2 and Group B3. Body conditions from Group B1 are substituted by similar versions of the average answer produced by the entire set of body conditions (both Group A and Group B). Elements of Group B2 are replaced

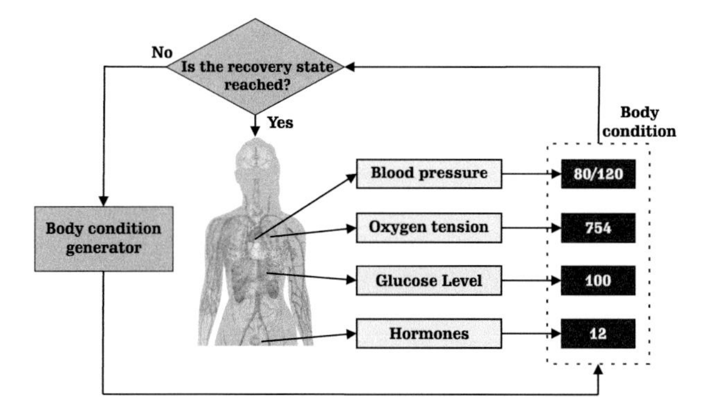

Fig. 5.1 Allostasis, a simple scheme

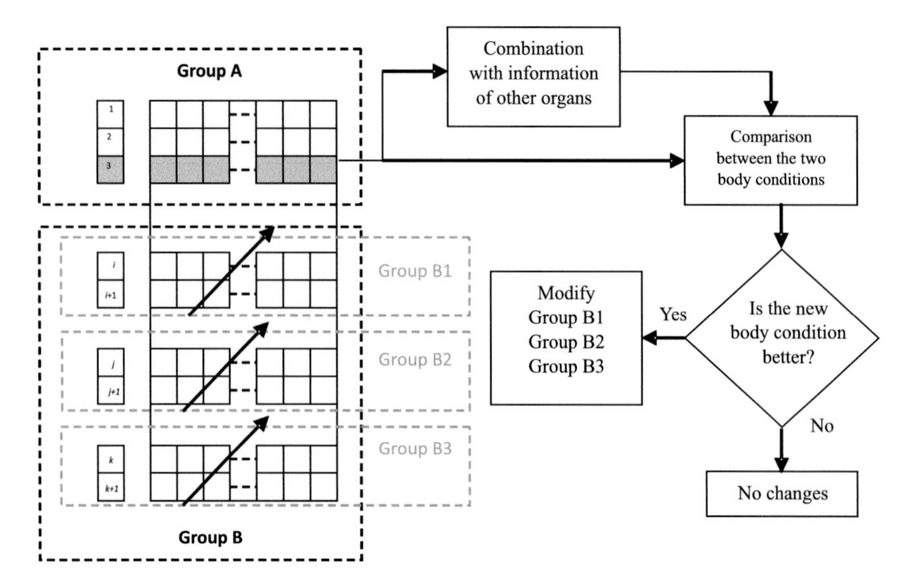

Fig. 5.2 Allostasis mechanism to generate new body conditions

by body conditions randomly generated inside the average answer. Finally, body conditions from Group B3 are replaced by body conditions which have demonstrated to be successful when a similar decompensation has happened. Figure 5.2 depicts graphically the allostasis mechanism to generate new body conditions.

It is important to remark that this explanation does not intend to be complete and precise, but only to give the necessary concepts in order to emulate the biological mechanism as evolutionary operators.

5.3 Allostatic Optimization (AO) Algorithm

The AO algorithm implements a set of operations that resembles the interaction rules that model the allostasis mechanism. In the approach, each solution within the search space represents a body condition. The "fitness value" evaluates the capacity of each body condition (individual) to reach a steady health state (good solution). The complete process mimics the allostasis phenomena.

The AO defines several operators. The combination operation is considered the main operator which is applied over all individuals of the population. Other operators called "collective" (operator group B1, B2 and B3) are also implemented in AO. Such operators affect only a group of elements.

Following the biological model of allostasis, the AO approach divides the entire population in four different groups: Group A, group B1, group B2 and group B3. The elements of group A are only modified by the combination operator whereas

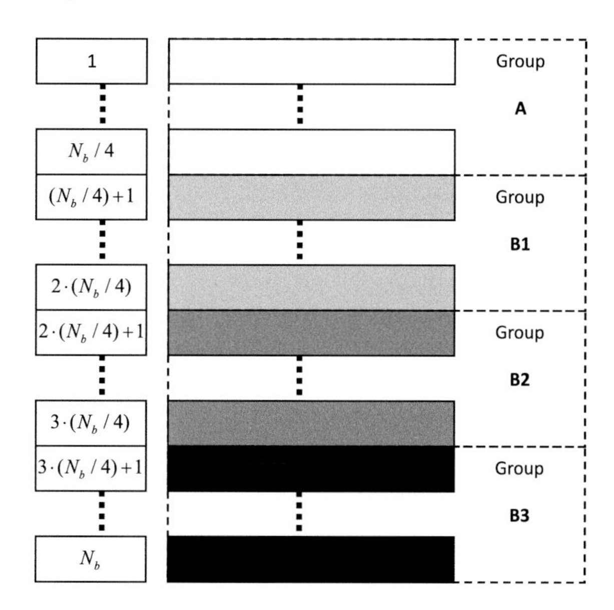

Fig. 5.3 Population distribution according to the AO approach

the elements of groups B1, B2 and B3 are affected by the combination operator and other collective operations. The size of each group is one fourth of the entire population. Thus, the population size must be selected in such a way that it can be entirely divided by four (20, 40, etc.). Figure 5.3 shows the population distribution according to the AO approach.

5.3.1 Description of the AO Algorithm

Similar to other evolutionary approaches, the AO algorithm is an iterative process that starts by initializing the population randomly (generated random solutions or body conditions). Then, the evolution process acts as following: The combination operator is applied to the first individual (body condition) of the population. As a result, a new individual is obtained. If the new individual is better than the original individual according to their fitness values, the original individual is replaced by the new one whereas the groups B1, B2 and B3 are modified used collective operators. However, if the original individual is better than the new individual, no changes are executed to the population. A complete iteration has lapsed when the combination operator has been applied to the last individual. This procedure is applied until a termination criterion is met (i.e. the iteration number *NI*).

According to the evolution process of AO, the following operators are employed:

1. Initialization.
2. Combination.

3. Collective B1.
4. Collective B2.
5. Collective B3.
6. Update the best element

Next, each operator is defined.

5.3.1.1 Initialization

The algorithm begins by initializing a set B of N_b body conditions ($\mathbf{B} = \{\mathbf{b}_1, \mathbf{b}_2, \ldots, \mathbf{b}_{N_b}\}$). Each body condition $\mathbf{b}_i$ is a D-dimensional vector containing parameter values to be optimized. Such values are randomly and uniformly distributed between the pre-specified lower initial parameter bound b_j^{low} and the upper initial parameter bound b_j^{high}.

$$\begin{aligned} b_{j,i} = b_j^{low} + \mathrm{rand}(0,1) \cdot (b_j^{high} - b_j^{low}), \\ i = 1, 2, \ldots, N_b; \quad j = 1, 2, \ldots, D \end{aligned} \tag{5.1}$$

with j and i being the parameter and individual indexes respectively. Hence, $b_{j,i}$ is the jth parameter of the ith individual. Once all body conditions were randomly generated, it is found the best individual $\mathbf{b}^{best}$ contained in the population, such that $\mathbf{b}^{best} \in \{\mathbf{B}\} | f(\mathbf{b}^{best}) = \max\{f(\mathbf{b}_1), f(\mathbf{b}_2), \ldots, f(\mathbf{b}_{N_b})\}$, where $f(\cdot)$ represents the objective function, part of the optimization problem to be solved (fitness function).

5.3.1.2 Combination

In allostasis, this operation combines each body condition with information provided by other organs. In this chapter, such effect is simulated by using a single mutation operation. Such operation replaces information of an original body condition with information extracted from other body condition. Therefore, the new individual combines information of both individuals. The combination operator uses as input a body condition $\mathbf{b}_i$ and produces as output a new body condition $\mathbf{b}_i^{new}$. In order to implement this operator, two different integers are randomly generated, Bc inside the number of body conditions $(1, 2, \ldots, N_b)$ and d inside the dimension number $(1, 2, \ldots, D)$. The combination takes place substituting the element $b_{d,i}$ from the original body condition $\mathbf{b}_i$ with the element $b_{d,Bc}$ from the body condition $\mathbf{b}_{Bc}$. Therefore, the only difference between the body conditions $\mathbf{b}_i$ and $\mathbf{b}_i^{new}$ is the element in the position d. Figure 5.4 shows graphically the combination operation.

The combination operator is sequentially applied beginning from the first body condition to the last one (N_b). Once the new individual $\mathbf{b}_i^{new}$ is generated by using the combination operator, it is compared, if the new individual $\mathbf{b}_i^{new}$ is better than

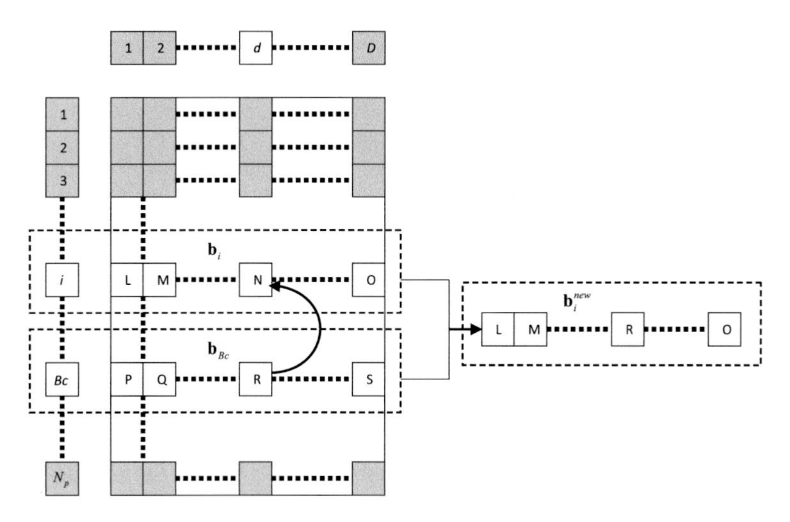

Fig. 5.4 The combination operator

the original individual $\mathbf{b}_i$ according to their fitness values. If $\mathbf{b}_i^{new}$ is better, $\mathbf{b}_i$ is replaced by $\mathbf{b}_i^{new}$ whereas the groups B1, B2 and B3 are modified used the collective operators. However, if $\mathbf{b}_i$ is better than $\mathbf{b}_i^{new}$, no changes are executed to the population.

5.3.1.3 Collective B1

This operator is executed only if $\mathbf{b}_i^{new}$ is better than the original $\mathbf{b}_i$. Such operator modifies only the elements of group B1, namely body conditions from $(N_b/4)+1$ to $2\cdot(N_b/4)$. In the allostasis mechanism, body conditions from Group B1are substituted by similar versions of the average answer produced by the entire set of body conditions. In the AO approach, the average answer $\mathbf{a} = \{a_1, a_2, \ldots, a_D\}$ is computed as:

$$a_j = \left(\frac{1}{N_b}\right)\cdot\sum_{i=1}^{N_b} b_{j,i}, \quad j = 1, 2, \ldots, D \tag{5.2}$$

The modification, applied to each element, depends on the existent difference between $\mathbf{b}_i^{new}$ and $\mathbf{b}^{best}$. Such relationship defined as m is calculated using:

$$m = \frac{1}{D}\cdot\left[\psi\cdot\left(1 - \frac{1}{e^{\psi\cdot\left[f(\mathbf{b}^{best}) - f(\mathbf{b}_i^{new})\right]}}\right)\right], \tag{5.3}$$

where $\psi \in [1, 3]$. Therefore, the body conditions of group B1 are updated according to:

$$b_{j,g1} = a_j - m + (2 \cdot m \cdot rand), \tag{5.4}$$

where $j \in \{1, 2, \ldots, D\}$, $g1 \in \{(N_b/4) + 1, (N_b/4) + 2, \ldots, 2 \cdot (N_b/4)\}$ and *rand* is a number randomly generated between 0 and 1.

5.3.1.4 Collective B2

According to the allostasis mechanism, elements of group B2 are replaced by body conditions randomly generated inside the average answer. Such effect is simulated modifying the elements of group B2 according to the following model:

$$b_{j,g2} = a_j \cdot rand, \tag{5.5}$$

where $g2 \in \{2 \cdot (N_b/4) + 1, 2 \cdot (N_b/4) + 2, \ldots, 3 \cdot (N_b/4)\}$.

5.3.1.5 Collective B3

Following the allostasis model, body conditions of Group B3 are replaced by body conditions which have demonstrated to be successful when a similar decompensation has happened. Such behavior is emulated producing perturbed versions of the best body condition $\mathbf{b}^{best} = \{b_1^{best}, b_2^{best}, \ldots, b_D^{best}\}$ found so-far. Thus, the elements of group B3 are modified by using:

$$b_{j,g3} = b_j^{best} - m + (2 \cdot m \cdot rand), \tag{5.6}$$

where $g3 \in \{3 \cdot (N_b/4) + 1, 3 \cdot (N_b/4) + 2, \ldots, N_b\}$.

5.3.1.6 Update the Best Element

In order to update de best body condition $\mathbf{b}^{best}$ seen so-far, the best found individual from the current population $\mathbf{b}^{best,k}$ is compared with the best individual $\mathbf{b}^{best,k-1}$ of the last generation. If $\mathbf{b}^{best,k}$ is better than $\mathbf{b}^{best,k-1}$ according to their fitness values, $\mathbf{b}^{best}$ is updated with $\mathbf{b}^{best,k}$, otherwise $\mathbf{b}^{best}$ remains without changes. Therefore, $\mathbf{b}^{best}$ stores the best historical body condition found so-far.

5.3.2 Discussion About the AO Algorithm

Evolutionary algorithms (EA) have been widely employed for solving complex optimization problems. These methods are found to be more powerful than conventional methods based on formal logics or mathematical programming [24]. In an

EA algorithm, search agents have to decide whether to explore unknown search positions or to exploit already tested positions in order to improve their solution quality. Pure exploration degrades the precision of the evolutionary process but increases its capacity to find new potentially solutions. On the other hand, pure exploitation allows refining existent solutions but adversely drives the process to local optimal solutions. Therefore, the ability of an EA to find a global optimal solution depends on its capacity to find a good balance between the exploitation of found-so-far elements and the exploration of the search space [25]. So far, the exploration–exploitation dilemma has been an unsolved issue within the framework of evolutionary algorithms.

Since the optimization point of view, the use of allostasis as metaphor introduces new concepts in the evolutionary algorithms: The fact of dividing the entire population in different regions and the employment of specialized operators applied selectively to each region. Using this architecture, it is possible to improve the balance between exploitation and exploration, conserving in the same population, individuals who only achieve exploration (body conditions of group A) and individuals that only verify exploitation (body conditions of group B3). Furthermore, the allostasis mechanism introduces the concept of average individual (average body condition). Since the evolution process of an EA concentrates the search procedure by exploring single positions, the search process is totally controlled by the individuals recently produced. Under these circumstances, promising new points could be discarded by the use of a new bad individual. The average individual (group B1 and B2) is used to partially control the search process according to the average performance of the population. Such mechanism acts as a filter which avoids that very god individuals or extremely bad individuals influence the search process.

DE, ABC and PSO are undoubtedly the most employed EA methods that use nature-inspired concepts in the optimization procedure. Unfortunately, they suffer from the premature convergence, particularly in multimodal problems. Premature convergence is produced by the strong influence of the best particle in the evolution process. Such particle is used by DE, ABC and PSO in the movement equations as a main individual in order to attract other particles. Under such conditions, the exploitation phase is privileged by allowing the evaluation of new search positions around the best individual. However, the exploration process is seriously damaged, avoiding searching in unexplored areas. Different to DE, ABC and PSO, the presented algorithm implements evolutionary operators that avoid concentrating most of the particles in only one position favoring the exploration process and eliminating the flaws related to the premature convergence.

5.4 Experimental Results

The parameters used by Allostatic Optimization were set up experimentally, which are shown in Table 5.1, together the corresponding to the algorithms to perform comparisons, i.e., Differential Evolution [26], ere it is suggested that the version

Table 5.1 Parameters used by algorithms used in comparisons

DE				PSO					ABC	AO	
Version	Pop size	F	Cr	Version	Swarm size	ϕ_1	ϕ_2	ω	Colony	Pop size	ψ
best/1/bin	50	0.9	0.5	CPSO	50	1.8	1.8	$k/maxiter$	50	20	1.3

best/1/bin is robust and converge to optimum when it is used with benchmark functions, whereas Particle Swarm Optimization is tuned with the parameters indicated in [27] and Artificial Bee Colony Optimization tuned as in [28]. In this work are taken either the maximum number of iterations as 1000 or an error of 1×10^{-90} as stopping criteria for each algorithm; also, each algorithm was run 25 times in a PC based on Intel Core i7-2600, with 8 GB in Ram and programmed in Matlab, version 7.13.0.564. The benchmark functions used in the experiments are shown in Table 5.2. For all the experiments, it is shown the complete result obtained, in order to demonstrate the real performance given by each algorithm when it is optimizing a determined function from benchmark (Table 5.3).

In the experimental part, a set corresponding to 21 benchmark functions was tested with Allostatic Optimization against DE, PSO and ABC; such benchmark was formed with functions with a fixed number of dimensions as well as functions of dimension of 6. The results of this experimental part are shown in Table 5.4 and the correspondent explanation is given below it.

Table 5.2 Unimodal test functions

Name	Function	Limits	f^*
Rosenbrock	$f_1(\mathbf{x}) = \sum_{i=1}^{D-1}\left[100(x_{i+1} - x_i^2)^2 - (x_i - 1)^2\right]$	$[-5, 10]^D$	0
Dixon&price	$f_2(\mathbf{x}) = (x_1 - 1)^2 + \sum_{i=2}^{D} i(x_i^2 - x_{i-1})^2$	$[-10, 10]^D$	0
Sphere	$f_3(\mathbf{x}) = \sum_{i=1}^{D} x_i^2$	$[-100, 100]^D$	0
Zakharov	$f_4(\mathbf{x}) = \sum_{i=1}^{D} x_i^2 + (\sum_{i=1}^{D} 0.5x_i)^2 + (\sum_{i=1}^{D} 0.5x_i)^4$	$[-5, 10]^D$	0
Sum squares	$f_5(\mathbf{x}) = \sum_{i=1}^{D} ix_i^2$	$[-10, 10]^D$	0
Powell	$f_6(\mathbf{x}) = \sum_{i=1}^{D/k} (x_{4i-3} + 10x_{4i-2})^2 + 5(x_{4i-1} - x_{4i})^2 \cdots + (x_{4i-2} - x_{4i-1})^4 + 10(x_{4i-3} - x_{4i})^4$	$[-4, 5]^D$	0
Quartic	$f_7(\mathbf{x}) = \sum_{i=1}^{D} ix_i^4 + random(0, 1)$	$[-1.28, 1.28]^D$	0
Step	$f_8(\mathbf{x}) = \sum_{i=1}^{D} (\lfloor x_i + 0.5 \rfloor)^2$	$[-100, 100]^D$	0
Schwefel 2.22	$f_9(\mathbf{x}) = \sum_{i=1}^{D} \lvert x_i \rvert + \prod_{i=1}^{D} \lvert x_i \rvert$	$[-10, 10]^D$	0
Schwefel 1.2	$f_{10}(\mathbf{x}) = \sum_{i=1}^{D} \left(\sum_{j=1}^{i} x_j\right)^2$	$[-100, 100]^D$	0
Schwefel 2.21	$f_{11}(\mathbf{x}) = \max_i\{\lvert x_i \rvert, 1 \le i \le D\}$	$[-100, 100]^D$	0
Rotated hyper-ellipsoid	$f_{12}(\mathbf{x}) = \sum_{i=1}^{D} \left(\sum_{j=1}^{i} x_j\right)^2$	$[-100, 100]^D$	0
Hyper-ellipsoid	$f_{13}(\mathbf{x}) = \sum_{i=1}^{D} i^2 x_i^2$	$[-1, 1]^D$	0

Table 5.3 Multimodal test functions

Name	Function	Limits	f^*
Ackley	$f_{14}(\mathbf{x}) = -20\exp\left(-0.2\sqrt{\frac{1}{D}\sum_{i=1}^{D} x_i^2}\right) - \exp\left(\frac{1}{D}\sum_{i=1}^{D}\cos(2\pi x_i)\right) + 20$	$[-15, 30]^D$	0
Griewank	$f_{15}(\mathbf{x}) = \frac{1}{4000}\sum_{i=1}^{D} x_i^2 - \prod_{i=1}^{D}\cos\left(\frac{x_i}{\sqrt{i}}\right) + 1$	$[-600, 600]^D$	0
Levy	$f_{16}(\mathbf{x}) = \sin^2(\pi y_0) + \sum_{i=1}^{D}(y_i - 1)^2\left(1 + 10\sin^2(\pi y_1 + 1)\right)\cdots$ $+ (y_{D-1} - 1)^2\left(1 + 10\sin^2(\pi y_{D-1} + 1)\right)$ $y_i = 1 + \frac{x_i + 1}{4}$	$[-10, 10]^D$	0
Schwefel	$f_{17}(\mathbf{x}) = \sum_{i=1}^{D} -x_i\sin\left(\sqrt{\lvert x_i\rvert}\right)$	$[-500, 500]^D$	-418.9828 * D
Rastrigin	$f_{18}(\mathbf{x}) = \sum_{i=1}^{D}\left[x_i^2 - 10\cos(2\pi x_i) + 10\right]$	$[-5.12, 5.12]^D$	0
Penalized 1	$f_{19}(\mathbf{x}) = \frac{\pi}{D}\left\{10\sin(\pi y_1) + \sum_{i=1}^{D-1}(y_i - 1)^2\left[1 + 10\sin^2(\pi y_{i+1})\right]\cdots\right.$ $\left. + (y_n - 1)^2\right\} + \sum_{i=1}^{D} u(x_i, 10, 100, 4)$ $y_i = 1 + \frac{(x_i + 1)}{4}, \quad u(x_i, a, k, m) = \begin{cases} k(x_i - a)^m & x_i > a \\ 0 & -a \le x_i \le a \\ k(-x_i - a)^m & x_i < a \end{cases}$	$[-50, 50]^D$	0
Penalized 2	$f_{20}(\mathbf{x}) = 0.1\left\{\sin^2(3\pi x_1) + \sum_{i=1}^{D}(x_i - 1)^2\left[1 + \sin^2(3\pi x_i + 1)\right]\cdots\right.$ $\left. + (x_n - 1)^2\left[1 + \sin^2(2\pi x_n)\right]\right\} + \sum_{i=1}^{D} u(x_i, 5, 100, 4)$	$[-50, 50]^D$	0
Salomon	$f_{21}(\mathbf{x}) = -\cos\left(2\pi\sqrt{\sum_{i=1}^{D} x_i^2}\right) + 0.1\sqrt{\sum_{i=1}^{D} x_i^2} + 1$	$[-100, 100]^D$	0
Alpine	$f_{22}(\mathbf{x}) = \sum_{i=1}^{D}\left(\lvert x_i\sin x_i\rvert + 0.1x_i\right)$	$[-10, 10]^D$	0

Table 5.4 Minimization of benchmark functions with d = 6 and the maximum number of evaluations of objective function it is 25,000

Function	AO	DE	PSO	ABC
	Mean (STD)	Mean (STD)	Mean (STD)	Mean (STD)
Rosenbrock	5.291e−001 (6.033e−001)	1.753e−002 (1.941e−002)	7.683e−001 (1.373e+000)	8.220e−002 (1.194e−001)
Dixon&price	3.114e−003 (5.133e−003)	3.176e−007 (5.210e−007)	3.733e−001 (3.377e−001)	1.389e−005 (1.770e−005)
Sphere	4.345e−003 (2.173e−002)	5.457e−079 (1.111e−078)	3.182e−039 (5.360e−039)	5.015e−017 (2.002e−017)
Zakharov	2.199e−002 (6.379e−002)	1.983e−003 (4.525e−003)	8.442e−032 (2.104e−031)	9.140e−003 (8.588e−003)
Sum squares	1.221e−018 (6.099e−018)	8.928e−080 (2.582e−079)	2.701e−041 (3.639e−041)	5.269e−017 (1.596e−017)
Powell	4.136e−002 (2.025e−001)	2.841e−006 (5.617e−006)	1.101e−007 (1.284e−007)	3.952e−004 (3.232e−004)
Quartic	6.734e−004 (7.214e−004)	1.135e−003 (5.694e−004)	1.033e−003 (7.766e−004)	5.274e−003 (2.476e−003)
Step	1.222e−015 (6.108e−015)	1.371e+000 (1.179e−001)	0.000e+000 (0.000e+000)	5.233e−017 (1.543e−017)
Schwefel 2.22	1.168e−020 (4.123e−020)	1.597e−040 (1.734e−040)	1.800e−021 (2.244e−021)	1.704e−016 (6.195e−017)
Schwefel 1.2	1.142e−013 (5.712e−013)	3.573e−078 (1.119e−077)	1.574e−038 (4.714e−038)	4.523e−017 (1.611e−017)
Schwefel 2.21	7.543e−003 (3.399e−002)	1.518e−011 (8.175e−012)	1.588e−017 (1.918e−017)	6.627e−004 (3.771e−004)
Rotated hyper-ellipsoid	1.106e−001 (6.823e−002)	6.503e+000 (1.173e+001)	2.855e−025 (8.388e−025)	1.444e+000 (1.240e+000)
Hyper-ellipsoid	6.06e−027 (1.17e−006)	3.74e−006 (7.33e−006)	0.803 (0.0164)	3.150e+003 (2.140e+003)
Ackley	2.8596e−001 (9.8972e−001)	4.441e−015 (0.000e+000)	4.157e−015 (9.837e−016)	4.638e−015 (8.253e−016)
Griewank	1.197e−002 (1.021e−002)	5.284e−003 (5.708e−003)	3.455e−002 (1.827e−002)	5.635e−003 (4.936e−003)
Levy	8.017e−009 (4.008e−008)	1.500e−032 (8.380e−048)	1.508e−032 (5.216e−048)	5.031e−017 (1.794e−017)
Schwefel	−2.516e+003 (6.622e+000)	−3.034e+003 (4.492e+002)	−2.234e+003 (1.967e+002)	−2.514e+003 (1.660e−004)
Rastrigin	5.478e−004 (2.739e−003)	0.000e+000 (0.000e+000)	9.552e−001 (1.130e+000)	0.000e+000 (0.000e+000)
Penalized 1	1.853e−007 (9.267e−007)	2.064e+000 (6.522e−001)	0.000e+000 (0.000e+000)	3.422e−017 (1.420e−017)
Penalized 2	3.374e−011 (1.687e−010)	0.000e+000 (0.000e+000)	0.000e+000 (0.000e+000)	4.333e−017 (1.607e−017)
Salomon	1.719e−001 (1.621e−001)	1.127e−001 (4.197e−002)	9.987e−002 (4.841e−017)	1.600e−001 (5.758e−002)
Alpine	7.719e−006 (8.871e−006)	6.994e−016 (1.075e−015)	4.330e−016 (6.127e−016)	4.171e−012 (9.745e−012)

Table 5.5 Minimization of benchmark functions with d = 30 and maximum number of objective function evaluations = 50,000

Function	AO	DE	PSO	ABC
	Mean (STD)	Mean (STD)	Mean (STD)	Mean (STD)
Rosenbrock	6.9377e+000 (1.9498e+001)	1.135e+000 (6.704e−001)	2.391e+003 (1.108e+004)	5.015e−001 (3.249e−001)
Dixon&Price	8.7871e−001 (1.2136e+000)	6.686e−002 (3.896e−002)	4.374e+001 (9.062e+001)	2.105e−002 (1.977e−002)
Sphere	2.3057e−028 (5.9027e−028)	1.626e−007 (6.983e−008)	2.696e−016 (4.587e−016)	1.431e−011 (1.406e−011)
Zakharov	7.2394e+001 (2.5326e+001)	3.229e+001 (1.252e+001)	1.495e+002 (1.032e+002)	2.776e+002 (4.433e+001)
Sum squares	2.8734e−030 (5.2109e−030)	2.207e−008 (8.367e−009)	3.200e+001 (9.883e+001)	2.194e−012 (2.665e−012)
Powell	1.5290e−001 (7.2728e−002)	1.985e−001 (1.018e−001)	5.751e+001 (6.387e+001)	5.162e−002 (1.763e−002)
Quartic	1.0857e−002 (6.0444e−003)	9.700e−002 (1.997e−002)	1.485e−002 (5.558e−003)	2.463e−001 (6.088e−002)
Step	1.0732e−028 (1.8666e−028)	7.289e+000 (6.856e−002)	5.438e−016 (1.235e−015)	9.311e−012 (6.243e−012)
Schwefel 2.22	7.1147e−016 (1.6339e−015)	1.325e−004 (3.034e−005)	1.600e+000 (3.742e+000)	8.778e−007 (3.948e−007)
Schwefel 1.2	2.9537e−027 (9.1499e−027)	2.371e−006 (1.386e−006)	8.000e+002 (2.769e+003)	1.520e−010 (1.370e−010)
Schwefel 2.21	9.7394e−001 (3.7040e−001)	3.817e+001 (3.190e+000)	9.610e−001 (5.650e−001)	4.568e+001 (7.009e+000)
Rotated hyper-ellipsoid	4.4837e+003 (1.4822e+003)	1.904e+004 (4.847e+003)	2.578e+003 (3.496e+003)	1.208e+004 (2.361e+003)
Hyper-ellipsoid	1.474e−007 (4.330e−007)	2.412e−009 (3.732e−009)	3.086e−003 (5.657e−003)	8.285e−010 (8.094e−010)
Ackley	3.8689e−014 (1.0250e−014)	2.165e−004 (5.697e−005)	2.527e+000 (3.937e+000)	7.209e−006 (5.063e−006)
Griewank	1.5716e−002 (7.0166e−002)	2.328e−003 (5.907e−003)	1.289e−002 (1.318e−002)	1.002e−003 (5.004e−003)
Levy	1.1723e−030 (2.6895e−030)	1.143e−013 (5.619e−014)	2.968e+000 (2.868e+000)	1.898e−013 (1.349e−013)
Schwefel	−1.2569e+004 (2.6996e−004)	−1.308e+004 (4.669e+002)	−9.578e+003 (6.610e+002)	−1.227e+004 (1.015e+002)
Rastrigin	8.4128e−014 (4.0594e−014)	1.592e−001 (3.723e−001)	7.073e+001 (1.773e+001)	2.216e−001 (3.887e−001)
Penalized 1	7.0450e−031 (1.3099e−030)	2.825e+000 (1.941e−001)	1.791e−001 (3.760e−001)	5.013e−013 (4.971e−013)
Penalized 2	6.3103e−030 (9.8231e−030)	3.347e−012 (1.846e−012)	3.308e−001 (7.292e−001)	8.755e−012 (2.110e−011)
Salomon	8.7987e−001 (1.9579e−001)	2.012e+000 (1.275e−001)	4.399e−001 (8.165e−002)	2.102e+000 (3.286e−001)
Alpine	7.1966e−003 (2.9934e−003)	1.804e−004 (1.263e−004)	1.776e−001 (8.880e−001)	7.699e−004 (8.896e−004)

Table 5.6 Minimization of benchmark functions with d = 50 and maximum number of objective function evaluations = 50,000

Function	AO	DE	PSO	ABC
	Mean (STD)	Mean (STD)	Mean (STD)	Mean (STD)
Rosenbrock	4.3901e+001 (4.5329e+001)	1.658e+001 (4.822e+000)	1.589e+004 (2.549e+004)	1.019e+001 (9.977e+000)
Dixon&Price	5.1390e+000 (3.0071e+000)	4.188e+000 (1.135e+000)	5.854e+003 (2.004e+004)	1.807e+000 (1.942e+000)
Sphere	3.4282e−014 (1.0509e−013)	7.450e−002 (1.887e−002)	4.000e+002 (2.000e+003)	2.322e−006 (1.824e−006)
Zakharov	1.7230e+002 (4.4007e+001)	8.770e+001 (2.236e+001)	7.210e+002 (2.124e+002)	5.942e+002 (4.503e+001)
Sum squares	4.1227e−015 (8.4765e−015)	1.645e−002 (4.614e−003)	3.640e+002 (4.572e+002)	1.097e−006 (8.860e−007)
Powell	1.0114e+000 (9.7937e−001)	1.450e+001 (8.185e+000)	1.992e+002 (4.600e+002)	2.893e−001 (1.743e−001)
Quartic	4.0305e−002 (1.9290e−002)	4.129e−001 (6.664e−002)	1.459e+000 (2.909e+000)	6.638e−001 (1.068e−001)
Step	5.0575e−014 (1.1607e−013)	1.391e+001 (2.217e−001)	5.170e−006 (1.223e−005)	2.903e−006 (3.267e−006)
Schwefel 2.22	1.3040e−008 (1.5132e−008)	1.207e−001 (1.583e−002)	5.613e+000 (7.672e+000)	1.496e−003 (4.906e−004)
Schwefel 1.2	2.9628e−013 (5.5100e−013)	1.428e+000 (2.663e−001)	2.600e+004 (2.517e+004)	4.223e−005 (1.081e−004)
Schwefel 2.21	1.1319e+001 (3.4154e+000)	7.003e+001 (3.798e+000)	2.122e+001 (3.455e+000)	7.315e+001 (4.950e+000)
Rotated hyper-ellipsoid	1.2844e+004 (3.9943e+003)	5.912e+004 (1.079e+004)	2.050e+004 (1.014e+004)	4.280e+004 (5.054e+003)
Hyper-ellipsoid	2.2534e+003 (5.8973e+003)	8.315e+003 (2.097e+004)	7.855e+001 (3.420e+001)	2.482e+002 (6.574e+003)
Ackley	2.4417e−008 (1.5623e−008)	1.523e−001 (3.312e−002)	5.066e+000 (3.513e+000)	8.913e−003 (4.861e−003)
Griewank	2.4223e−002 (5.5876e−002)	1.546e−001 (4.911e−002)	2.609e−002 (3.222e−002)	9.175e−004 (1.692e−003)
Levy	3.0487e−016 (8.5590e−016)	2.780e−006 (9.550e−007)	1.443e+001 (5.970e+000)	3.213e−007 (2.674e−007)
Schwefel	−2.0949e+004 (4.3091e−001)	−2.131e+004 (3.878e+002)	−1.434e+004 (9.221e+002)	−1.939e+004 (2.585e+002)
Rastrigin	1.9984e−008 (9.5951e−008)	3.752e+000 (1.193e+000)	1.606e+002 (2.740e+001)	7.038e+000 (2.698e+000)
Penalized 1	4.0600e−016 (1.4963e−015)	3.167e+000 (1.256e−001)	4.283e−001 (5.209e−001)	3.764e−007 (3.971e−007)
Penalized 2	1.9188e−015 (3.0588e−015)	7.730e−005 (2.658e−005)	5.136e+000 (4.090e+000)	2.296e−006 (3.480e−006)
Salomon	1.6799e+000 (2.4831e−001)	6.995e+000 (4.050e−001)	1.208e+000 (1.883e+000)	5.165e+000 (7.513e−001)
Alpine	3.3938e−002 (2.1200e−002)	1.806e−001 (6.899e−002)	5.098e+000 (5.228e+000)	6.562e−002 (4.859e−002)

Table 5.7 Minimization of benchmark functions with d = 100 and maximum number of objective function evaluations = 50,000

Function	AO	DE	PSO	ABC
	Mean (STD)	Mean (STD)	Mean (STD)	Mean (STD)
Rosenbrock	1.7433e+002 (7.3465e+001)	2.840e+003 (3.913e+002)	1.194e+005 (7.987e+004)	1.522e+002 (5.616e+001)
Dixon&Price	4.1898e+001 (1.0727e+001)	3.409e+003 (5.858e+002)	2.121e+005 (3.022e+005)	3.148e+001 (1.087e+001)
Sphere	2.3171e−004 (2.9524e−004)	8.536e+002 (1.937e+002)	5.273e+003 (6.564e+003)	5.720e−002 (1.157e−001)
Zakharov	3.9807e+002 (5.7117e+001)	3.061e+002 (4.203e+001)	2.669e+003 (4.455e+002)	1.367e+003 (5.652e+001)
Sum squares	5.1558e−005 (5.7525e−005)	3.708e+002 (5.092e+001)	4.502e+003 (3.475e+003)	1.157e−002 (1.686e−002)
Powell	5.9186e+000 (3.6590e+000)	5.816e+002 (7.572e+001)	2.294e+003 (1.651e+003)	2.407e+001 (1.914e+001)
Quartic	1.8949e−001 (5.0867e−002)	3.001e+000 (2.992e−001)	1.830e+001 (2.609e+001)	2.363e+000 (3.190e−001)
Step	1.8702e−004 (2.2635e−004)	1.094e+003 (1.444e+002)	4.548e+003 (7.057e+003)	2.419e−002 (2.783e−002)
Schwefel 2.22	5.1005e−003 (2.4142e−003)	1.957e+001 (1.477e+000)	3.964e+001 (1.610e+001)	8.807e−002 (1.715e−002)
Schwefel 1.2	5.8380e−003 (4.2684e−003)	3.961e+004 (7.158e+003)	3.954e+005 (3.277e+005)	8.464e−001 (5.346e−001)
Schwefel 2.21	3.2812e+001 (3.4286e+000)	8.851e+001 (1.253e+000)	5.416e+001 (3.545e+000)	9.035e+001 (1.290e+000)
Rotated hyper-ellipsoid	5.6813e+004 (1.2001e+004)	2.501e+005 (4.020e+004)	1.123e+005 (1.795e+004)	1.756e+005 (1.738e+004)
Hyper-ellipsoid	8.3244e+004 (2.5894e+004)	3.6597e+005 (9.654e+004)	2.2856e+005 (3.9875e+004)	3.6978e+005 (7.6598e+004)
Ackley	2.6608e−003 (1.4637e−003)	5.244e+000 (1.594e−001)	1.100e+001 (2.184e+000)	2.458e+000 (3.245e−001)
Griewank	1.7765e−002 (3.0909e−002)	8.492e+000 (8.483e−001)	4.468e+001 (5.876e+001)	1.786e−001 (1.654e−001)
Levy	2.6073e−006 (3.0943e−006)	4.884e−001 (1.053e−001)	7.008e+001 (1.878e+001)	7.960e−002 (1.658e−001)
Schwefel	−4.1441e+004 (2.2116e+002)	−4.241e+004 (5.314e+002)	−2.648e+004 (2.040e+003)	−3.490e+004 (5.280e+002)
Rastrigin	1.8081e+000 (1.3318e+000)	2.271e+002 (1.118e+001)	4.639e+002 (7.672e+001)	6.899e+001 (1.174e+001)
Penalized 1	6.3448e−006 (2.8778e−005)	9.113e+000 (7.392e−001)	1.937e+001 (1.080e+001)	2.859e−003 (8.804e−003)
Penalized 2	1.8451e−005 (1.6437e−005)	2.519e+001 (2.617e+001)	2.822e+002 (3.817e+002)	4.386e−003 (7.684e−003)
Salomon	4.9233e+000 (4.1390e−001)	2.846e+001 (8.118e−001)	7.983e+000 (4.486e+000)	1.883e+001 (1.920e+000)
Alpine	3.7172e−001 (1.3017e−001)	2.213e+001 (1.849e+000)	1.857e+001 (9.103e+000)	1.943e+000 (6.843e−001)

As it can be easily noticed, when the problem is given with low dimensionality, AO is easily outperformed by DE, PSO and ABC, whose behavior it is almost the same, and even they perform equal among them in some functions. However, as the dimension of the problem increases, the performance of those algorithms it is turned in a different way, as can be noticed in Tables 5.5, 5.6 and 5.7, where were eliminated the functions with fix dimension. Besides, in Fig. 5.5 it is shown a closer look to the evolution of the four algorithms when they are looking for the optima in "Quartic" function for 30, 50 and 100 dimensions, and also it is shown the 2D version of such function in Fig. 5.6.

A closer look to the "Schwefel" function shows that DE apparently has better performance than AO; however, the result given by DE it is outside the range given for the optima, which is −418.9828 * d. Therefore, we consider that AO evolve to an adequate result for aforementioned function. As dimension increases, the PSO algorithm has the worst efficiency of all compared algorithm. Even though that optima in "Zakharov" function is not reached for 30, 50 and 100 dimensions, in those experiments AO gives the second best result, while the best it is given by DE. The same situation applies for the "Powell" function in 30 and 50 dimensions.

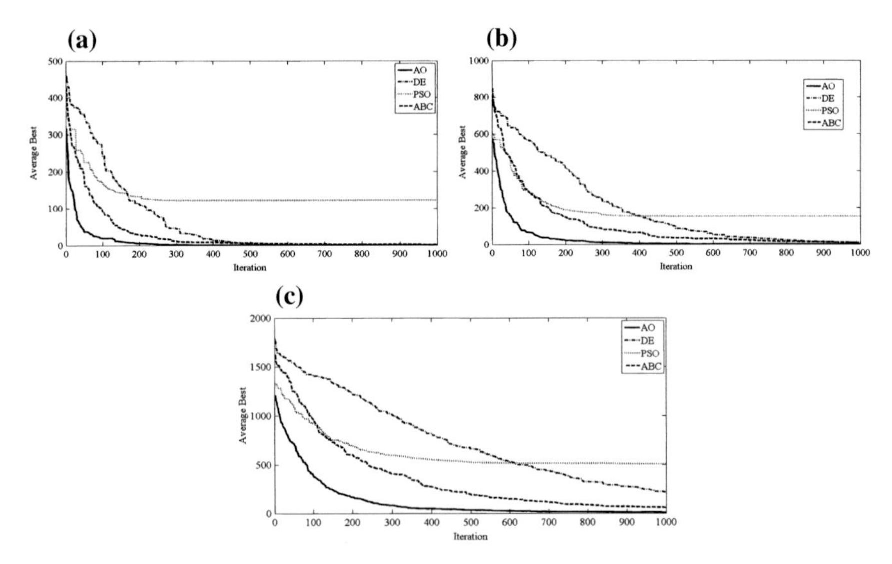

Fig. 5.5 Evolution of AO, DE, PSO and ABC over the "Rastrigin" function. **a** 30 dimensions. **b** 50 dimensions. **c** 100 dimensions

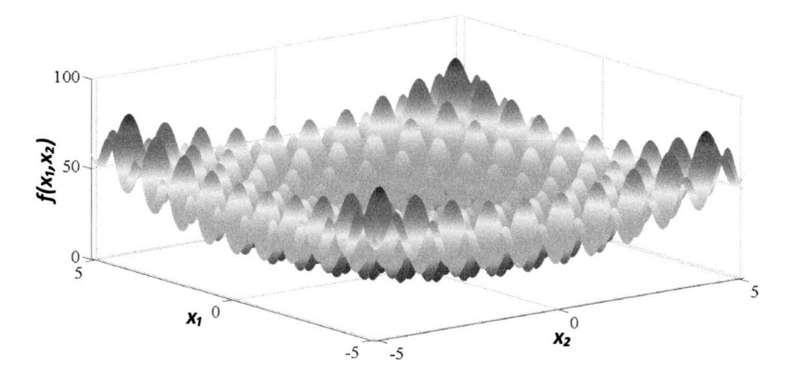

Fig. 5.6 A representation in 2D of "Rastrigin" function

5.5 Summary

In this work it is presented a new bio-inspired method that mimics the way in which human body cope with stress, by adjusting accordingly homeostatic systems in a coordinated fashion. We call this metaheuristic Allostatic Optimization. In this algorithm, the agents modify their homeostatic systems in order to cope with artificial stress; such changes appertain to movements into the search space, and a good response to stress corresponds to a good fitness found. Also, it is used a single modification of individuals to simulate artificial stress, and the only way in which the individuals in population have communication it is through a homeostatic collective memory, which it is constantly updated with the best homeostatic movements done by a particular individual. According to the empirical results obtained through a series of experiments, we conclude than the proposal it is at least similar in behavior than two state of the art metaheuristics called Differential Evolution and Artificial Bee Colony optimization. On the other hand, Allostatic Optimization outperforms Particle Swarm Optimization over a benchmark of 21 functions used to probe the new proposal of evolutionary methods. Future work includes comparisons among the proposal in other areas where problems to solve could be represented as optimization ones, as vision by computer.

References

1. Panos, M.P., Edwin, H.R., Tuy, H.: Recent developments and trends in global optimization. J. Comput. Appl. Math. **124**(1–2), 209–228 (2000)
2. Floudas, C., Akrotirianakis, I., Caratzoulas, S., Meyer, C., Kallrath, J.: Global optimization in the 21st century: advances and challenges. Comput. Chem. Eng. **29**(6), 1185–1202 (2005)
3. Ying, J., Ke-Cun, Z., Shao-Jian, Q.: A deterministic global optimization algorithm. Appl. Math. Comput. **185**(1), 382–387 (2007)

4. Lera, D., Sergeyev, Y.: Lipchitz and Hölder global optimization using space-filling curves. Appl. Numer. Math. **60**(1–2), 115–129 (2010)
5. Georgieva, A., Jordanov, I.: Global optimization based on novel heuristics, low-discrepancy sequences and genetic algorithms. Eur. J. Oper. Res. **196**(2), 413–422 (2009)
6. Kennedy, J., Eberhart, R.C.: Particle swarm optimization. In: Proceedings of the IEEE International Conference on Neural Networks, vol. 4, pp. 1942–1948 (1995)
7. Karaboga, D.: An idea based on honey bee swarm for numerical optimization. Technical report, TR06, Erciyes University, Engineering Faculty, Computer Engineering Department (2005)
8. Dorigo, M., Maniezzo, V., Colorni, A.: Positive feedback as a search strategy. Technical Report No. 91-016, Politecnico di Milano (1991)
9. Kirkpatrick, S., Gelatt, C., Vecchi, M.: Optimization by simulated annealing. Science **220** (4598), 671–680 (1983)
10. İlker, B., Birbil, S., Shu-Cherng, F.: An electromagnetism-like mechanism for global optimization. J. Global Optim. **25**(3), 263–282 (2003)
11. Rashedia, E., Nezamabadi-pour, H., Saryazdi, S.: 'Filter modeling using gravitational search algorithm. Eng. Appl. Artif. Intell. **24**(1), 117–122 (2011)
12. Geem, Z.W., Kim, J.H., Loganathan, G.V.: A new heuristic optimization algorithm: harmony search. Simulation **76**(2), 60–68 (2001)
13. Fogel, L.J., Owens, A.J., Walsh, M.J.: Artificial Intelligence through Simulated Evolution. Wiley, Chichester (1966)
14. De Jong, K.: Analysis of the Behavior of a class of genetic adaptive systems. PhD thesis, University of Michigan, Ann Arbor (1975)
15. Koza, J.R.: Genetic programming: a paradigm for genetically breeding populations of computer programs to solve problems. Rep. No. STAN-CS-90-1314, Stanford University, CA (1990)
16. Holland, J.H.: Adaptation in Natural and Artificial Systems. University of Michigan Press, Ann Arbor (1975)
17. de Castro, L.N., Von Zuben, F.J.: Artificial immune systems: part I—basic theory and applications. Technical report, TR-DCA 01/99, December (1999)
18. Storn, R., Price, K.: Differential evolution—a simple and efficient heuristic for global optimization over continuous spaces. J. Global Optim. **11**(4), 341–359 (1995)
19. Norouzzadeh, M.S., Ahmadzadeh, M.R., Palhang, M.: LADPSO: using fuzzy logic to conduct PSO algorithm. Appl. Intell. **37**(2), 290–304 (2012)
20. Ali, Y.M.B.: Psychological model of particle swarm optimization based multiple emotions. Appl. Intell. **36**(3), 649–663 (2012)
21. Cannon, W.B.: Bodily changes in pain, hunger, fear and rage: an account of recent researchers into the function of emotional excitement, 2nd edn. Appleton, New York (1929)
22. Cannon, W.B.: The Wisdom of the Body. W.W. Norton, New York (1932)
23. Gross, C.G.: Claude Bernard and the constancy of the internal environment. Neuroscientist **4**, 380–385 (1988)
24. Yang, X.-S.: Nature-inspired metaheuristic algorithms. Luniver Press, Beckington (2008)
25. Chen, D.B., Zhao, C.X.: Particle swarm optimization with adaptive population size and its application. Appl. Soft Comput. **9**(1), 39–48 (2009)
26. Mezura-Montes, E., Velázquez-Reyes, J., Carlos, A., Coello Coello, A.: comparative study of differential evolution variants for global optimization. In: Proceedings of the 8th annual conference on Genetic and evolutionary computation (GECCO '06). ACM, New York, NY, USA, pp. 485–492 (2006)
27. Vesterstrom, J., Thomsen, R.: A comparative study of differential evolution, particle swarm optimization, and evolutionary algorithms on numerical benchmark problems, Evolutionary Computation, 2004. CEC2004. Congress on, vol. 2, pp. 1980–1987, 19–23 June 2004
28. Karaboga, D., Akay, B.: A comparative study of Artificial Bee Colony algorithm, Appl. Math. Comput. **214**(1), 108–132. ISSN 0096-3003, 1 Aug 2009

Chapter 6
Optimization Based on the Behavior of Locust Swarms

6.1 Introduction

The collective intelligent behavior of insect or animal groups in nature such as flocks of birds, colonies of ants, schools of fish, swarms of bees and termites have all attracted the attention of researchers. Despite single members of swarms are non-sophisticated individuals, they are able to achieve complex tasks in cooperation. The collective swarm behavior emerges from relatively simple actions or interactions among the members. Entomologists have studied this collective phenomenon to model biological swarms while engineers have applied these models as a framework for solving complex real-world problems. The discipline of artificial intelligence which is concerned with the design of intelligent multi-agent algorithms by taking inspiration from the collective behavior of social insects or animals is known as swarm intelligence [1]. Swarm algorithms have several advantages such as scalability, fault tolerance, adaptation, speed, modularity, autonomy and parallelism [2].

Several swarm algorithms have been developed by a combination of deterministic rules and randomness, mimicking the behavior of insect or animal groups in nature. Such methods include the social behavior of bird flocking and fish schooling such as the Particle Swarm Optimization (PSO) algorithm [3], the cooperative behavior of bee colonies such as the Artificial Bee Colony (ABC) technique [4], the social foraging behavior of bacteria such as the Bacterial Foraging Optimization Algorithm (BFOA) [5], the simulation of the herding behavior of krill individuals such as the Krill Herd (KH) method [6], the mating behavior of firefly insects such as the Firefly (FF) method [7], the emulation of the lifestyle of cuckoo birds such as the Cuckoo Search (CS) [8], the social-spider behavior such as the Social Spider Optimization (SSO) [9], the simulation of the animal behavior in a group such as the Collective Animal Behavior (CAB) [10] and the emulation of the differential evolution in species such as the Differential Evolution (DE) [11].

In particular, insect swarms and animal groups provide a rich set of metaphors for designing swarm optimization algorithms. Such methods are complex systems

E. Cuevas et al., *Advances of Evolutionary Computation: Methods and Operators*, Studies in Computational Intelligence 629,
DOI 10.1007/978-3-319-28503-0_6

composed by individuals that tend to reproduce specialized behaviors [12]. However, most of swarm algorithms and other evolutionary algorithms tend to exclusively concentrate individuals in the current best positions. Under such circumstances, these algorithms seriously limit their search capacities.

Although PSO and DE are the most popular algorithms for solving complex optimization problems, they present serious flaws such as premature convergence and difficulty to overcome local minima [13, 14]. The cause for such problems is associated to the operators that modify individual positions. In such algorithms, during their evolution, the position of each agent for the next iteration is updated yielding an attraction towards the position of the best particle seen so-far (in case of PSO) or towards other promising individuals (in case of DE). As the algorithm evolves, these behaviors cause that the entire population rapidly concentrates around the best particles, favoring the premature convergence and damaging the appropriate exploration of the search space [15, 16].

The interesting and exotic collective behavior of insects have fascinated and attracted researchers for many years. The intelligent behavior observed in these groups provides survival advantages, where insect aggregations of relatively simple and "unintelligent" individuals can accomplish very complex tasks using only limited local information and simple rules of behavior [17]. Locusts (Schistocerca Gregaria) are a representative example of such collaborative insects [18]. Locust is a kind of grasshopper that can change reversibly between a solitary and a social phase, with clear behavioral differences among both phases [19]. The two phases show many differences regarding the overall level of activity and the degree to which locusts are attracted or repulsed among them [20]. In the solitary phase, locusts avoid contact to each other (locust concentrations). As consequence, they distribute throughout the space, exploring sufficiently over the plantation [20]. On other hand, in the social phase, locusts frantically concentrate around those elements that have already found good food-sources [21]. Under such a behavior, locust attempt to efficiently find better nutrients by devastating promising areas within the plantation.

In this chapter, a novel swarm algorithm, called the Locust Search (LS) is presented for solving some optimization tasks. The LS algorithm is based on the behavioral simulation of swarms of locusts. In the presented algorithm, individuals emulate a group of locusts which interact to each other based on the biological laws of the cooperative swarm. The algorithm considers two different behaviors: solitary and social. Depending on the behavior, each individual is conducted by a set of evolutionary operators which mimics different cooperative conducts that are typically found in the swarm. Different to most of existent swarm algorithms, the behavioral model in the presented approach explicitly avoids the concentration of individuals in the current best positions. Such fact allows not only to emulate in a better realistic way the cooperative behavior of the locust colony, but also to incorporate a computational mechanism to avoid critical flaws that are commonly present in the popular PSO and DE algorithms, such as the premature convergence and the incorrect exploration-exploitation balance. In order to illustrate the proficiency and robustness of the presented approach, its performance is compared to other well-known evolutionary methods. The comparison examines several standard benchmark functions which are commonly considered in the literature. The results demonstrate a high performance of the presented method in the search of a global optimum for several benchmark functions.

This chapter is organized as follows. In Sect. 6.2, we introduce basic biological issues of the algorithm analogy. In Sect. 6.3, the novel LS algorithm and its characteristics are both described. Section 6.4 presents the experimental results and the comparative study, while Sect. 6.5 presents some concluding remarks.

6.2 Biological Fundamentals and Mathematical Models

Social insect societies are complex cooperative systems that self-organize within a set of constraints. Cooperative groups are good at manipulating and exploiting their environment, defending resources and brooding, yet allowing task specialization among group members [22, 23]. A social insect colony functions as an integrated unit that not only possesses the ability to operate at a distributed manner, but also undertakes a huge construction of global projects [24]. It is important to acknowledge that global order for insects can arise as a result of internal interactions among members.

Locusts are a kind of grasshoppers that exhibit two opposite behavioral phases: solitary and social (gregarious). Individuals in the solitary phase avoid contact to each other (locust concentrations). As consequence, they distribute throughout the space while sufficiently exploring the plantation [20]. In contrast, locusts in the gregarious phase gather into several concentrations. Such congregations may contain up to 10^{10} members, cover cross-sectional areas of up to 10 km^2, and a travelling capacity up to 10 km per day for a period of days or weeks as they feed causing a devastating crop loss [25]. The mechanism to switch from the solitary phase to the gregarious phase is complex and has been a subject of significant biological inquiry. Recently, a set of factors has been implicated to include geometry of the vegetation landscape and the olfactory stimulus [26].

Only few works [20, 21] that mathematically model the locust behavior have been published. Both approaches develop two different minimal models with the goal of reproducing the macroscopic structure and motion of a group of locusts. Considering that the method in [20] focuses on modelling the behavior for each locus in the group, its fundamentals have been employed to develop the algorithm that is presented in this chapter.

6.2.1 *Solitary Phase*

This sections describes how each locust's position is modified as a result of its behavior under the solitary phase. Considering that $\mathbf{x}_i^k$ represents the current position of the ith locust in a group of N different elements, the new position $\mathbf{x}_i^{k+1}$ is calculated by using the following model:

$$\mathbf{x}_i^{k+1} = \mathbf{x}_i^k + \Delta\mathbf{x}_i \tag{6.1}$$

with $\Delta\mathbf{x}_i$ corresponding to the change of position that is experimented by $\mathbf{x}_i^k$ as a consequence of its social interaction with all other elements in the group.

Two locusts in the solitary phase exert forces on each other according to basic biological principles of attraction and repulsion (see, e.g., [20]). Repulsion operates quite strongly over a short length scale in order to avoid concentrations. Attraction is weaker, and operates over a longer length scale, providing the social force that is required to maintain the group′s cohesion. Therefore, the strength of such social forces can be modeled by the following function:

$$s(r) = F \cdot e^{-r/L} - e^{-r} \tag{6.2}$$

Here, r is a distance, F describes the strength of attraction, and L is the typical attractive length scale. We have scaled the time and the space coordinates so that the repulsive strength and length scale are both represented by the unity. We assume that $F < 1$ and $L > 1$ so that repulsion is stronger and features in a shorter-scale, while attraction is applied in a weaker and longer-scale; both facts are typical for social organisms [21]. The social force exerted by locust j over locust i is:

$$\mathbf{s}_{ij} = s(r_{ij}) \cdot \mathbf{d}_{ij} \tag{6.3}$$

where $r_{ij} = |\mathbf{x}_j - \mathbf{x}_i|$ is the distance between the two locusts and $\mathbf{d}_{ij} = (\mathbf{x}_j - \mathbf{x}_i)/r_{ij}$ is the unit vector pointing from $\mathbf{x}_i$ to $\mathbf{x}_j$. The total social force on each locust can be modeled as the superposition of all of the pairwise interactions:

$$\mathbf{S}_i = \sum_{\substack{j=1 \\ j \neq i}}^{N} \mathbf{s}_{ij} \tag{6.4}$$

The change of position $\Delta\mathbf{x}_i$ is modeled as the total social force experimented by $\mathbf{x}_i^k$ as the superposition of all of the pairwise interactions. Therefore, $\Delta\mathbf{x}_i$ is defined as follows:

$$\Delta\mathbf{x}_i = \mathbf{S}_i \tag{6.5}$$

In order to illustrate the behavioral model under the solitary phase, Fig. 6.1 presents an example, assuming a population of three different members ($N = 3$) which adopt a determined configuration in the current iteration k. As a consequence of the social forces, each element suffers an attraction or repulsion to other elements

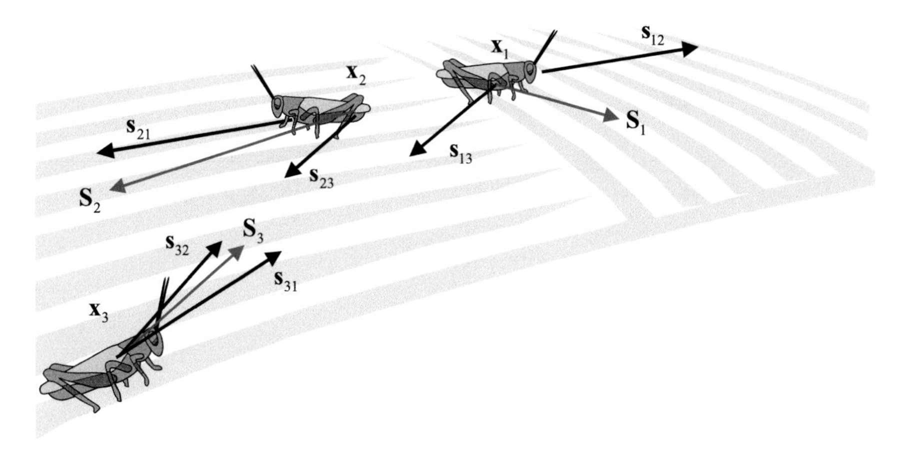

Fig. 6.1 Behavioral model under the solitary phase

depending on the distance among them. Such forces are represented by $\mathbf{s}_{12}, \mathbf{s}_{13}, \mathbf{s}_{21}, \mathbf{s}_{23}, \mathbf{s}_{31}, \mathbf{s}_{32}$. Since $\mathbf{x}_1$ and $\mathbf{x}_2$ are too close, the social forces $\mathbf{s}_{12}$ and $\mathbf{s}_{13}$ present a repulsive nature. On the other hand, as the distances $|\mathbf{x}_1 - \mathbf{x}_3|$ and $|\mathbf{x}_2 - \mathbf{x}_3|$ are quite long, the social forces $\mathbf{s}_{13}$, $\mathbf{s}_{23}$, $\mathbf{s}_{31}$ and $\mathbf{s}_{32}$ between $\mathbf{x}_1 \leftrightarrow \mathbf{x}_3$ and $\mathbf{x}_2 \leftrightarrow \mathbf{x}_3$, all belong to the attractive nature. Therefore, the change of position $\Delta\mathbf{x}_1$ is computed as the vector resultant between $\mathbf{s}_{12}$ and $\mathbf{s}_{13}(\Delta\mathbf{x}_1 = \mathbf{s}_{12} + \mathbf{s}_{13})$. The values $\Delta\mathbf{x}_2$ and $\Delta\mathbf{x}_3$ are also calculated accordingly.

In addition to the presented model [20], some studies [27–29] suggest that the social force $\mathbf{s}_{\mathrm{ij}}$ is also affected by the dominance of the involved individuals $\mathbf{x}_i$ and $\mathbf{x}_j$ in the pairwise process. Dominance is a property that relatively qualifies the capacity of an individual to survive, in relation to other elements in a group. The locust's dominance is determined by several characteristics such as size, chemical emissions, location with regard to food sources, etc. Under such circumstances, the social force is magnified or weakened depending on the most dominant individual that is involved in the repulsion-attraction process.

6.2.2 *Social Phase*

In this phase, locusts frantically concentrate around the elements that have already found good food sources. They attempt to efficiently find better nutrients by devastating promising areas within the plantation. In order to simulate the social phase, the food quality index Fq_i is assigned to each locust $\mathbf{x}_i$ of the group as such index reflexes the quality of the food source where $\mathbf{x}_i$ is located.

Under this behavioral model, each of the N elements of the group are ranked according to its corresponding food quality index. Afterwards, the b elements featuring the best food quality indexes are selected ($b \ll N$). Considering a

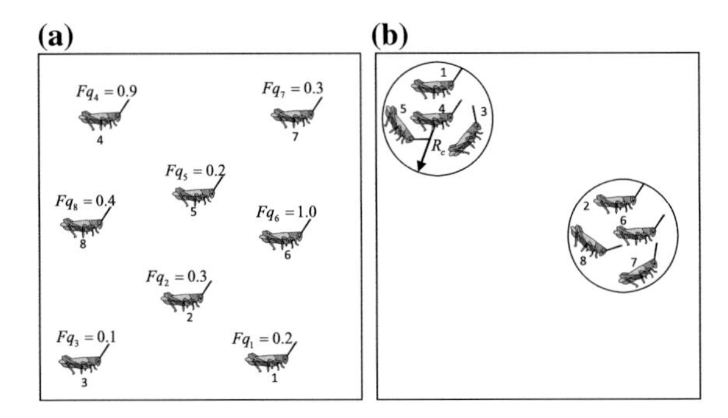

Fig. 6.2 Behavioral model under the social phase. **a** Initial configuration and food quality indexes. **b** Final configuration after the operation of the social phase

concentration radius R_c that is created around each selected element, a set of c new locusts is randomly generated inside R_c. As a result, most of the locusts will be concentrated around the best b elements. Figure 6.2 shows a simple example of the behavioral model under the social phase. In the example, the configuration includes eight locust ($N = 8$), just as it is illustrated by Fig. 6.2a that also presents the food quality index for each locust. A food quality index near to one indicates a better food source. Therefore, Fig. 6.2b presents the final configuration after the social phase, assuming $b = 2$.

6.3 The Locust Search (LS) Algorithm

In this chapter, some behavioral principles drawn from a swarm of locusts have been used as guidelines for developing a new swarm optimization algorithm. The LS assumes that entire search space is a plantation, where all the locusts interact to each other. In the presented approach, each solution within the search space represents a locust position inside the plantation. Every locust receives a food quality index according to the fitness value of the solution that is symbolized by the locust's position. As it has been previously discussed, the algorithm implements two different behavioral schemes: solitary and social. Depending on each behavioral phase, each individual is conducted by a set of evolutionary operators which mimics the different cooperative operations that are typically found in the swarm.

From the implementation point of view, in the LS operation, a population $\mathbf{L}^k(\{\mathbf{l}_1^k, \mathbf{l}_2^k, \ldots, \mathbf{l}_N^k\})$ of N locusts (individuals) is evolved from the initial point ($k = 0$) to a total gen number of iterations ($k = gen$). Each locust $\mathbf{l}_i^k$ ($i \in [1, \ldots, N]$) represents an n-dimensional vector $\left\{l_{i,1}^k, l_{i,2}^k, \ldots, l_{i,n}^k\right\}$ where each dimension corresponds to a decision variable of the optimization problem to be solved. The set of decision variables constitutes the feasible search space

$\mathbf{S} = \left\{ \mathbf{l}_i^k \in \mathbb{R}^n \middle| lb_d \leq l_{i,d}^k \leq ub_d \right\}$, where lb_d and ub_d corresponds to the lower and upper bounds for the dimension d, respectively. The food quality index that is associated to each locust $\mathbf{l}_i^k$ (candidate solution) is evaluated through an objective function $f(\mathbf{l}_i^k)$ whose final result represents the fitness value of $\mathbf{l}_i^k$. In the LS algorithm, each iteration of the evolution process consists of two operators: (A) solitary and (B) social. Beginning by the solitary stage, the set of locusts is operated in order to sufficiently explore the search space. On the other hand, the social operation refines existent solutions within a determined neighborhood (exploitation) subspace.

6.3.1 Solitary Operation (A)

One of the most interesting features of the presented method is the use of the solitary operator to modify the current locust positions. Under this approach, locusts are displaced as a consequence of the social forces produced by the positional relations among the elements of the swarm. Therefore, near individuals tend to repel each other, avoiding the concentration of elements in regions. On the other hand, distant individuals tend to attract to each other, maintaining the cohesion of the swarm. A clear difference to the original model in [20] considers that social forces are also magnified or weakened depending on the best fitness value (the most dominant) of the individuals that are involved in the repulsion-attraction process.

In the solitary phase, a new position $\mathbf{p}_i (i \in [1, \ldots, N])$ is produced by perturbing the current locust position $\mathbf{l}_i^k$ with a change of position $\Delta\mathbf{l}_i$ $(\mathbf{p}_i = \mathbf{l}_i^k + \Delta\mathbf{l}_i)$. The change of position $\Delta\mathbf{l}_i$ is the result of the social interactions experimented by $\mathbf{l}_i^k$ as a consequence of its repulsion-attraction behavioral model. Such social interactions are pairwise computed among $\mathbf{l}_i^k$ and the other $N - 1$ individuals in the swarm. In the original model, social forces are calculated by using Eq. 6.3. However, in the presented method, it is modified to include the best fitness value (the most dominant) of the individuals involved in the repulsion-attraction process. Therefore, the social force, that is exerted between $\mathbf{l}_j^k$ and $\mathbf{l}_i^k$, is calculated by using the following new model:

$$\mathbf{s}_{ij}^m = \rho(\mathbf{l}_i^k, \mathbf{l}_j^k) \cdot s(r_{ij}) \cdot \mathbf{d}_{ij} + rand(1, -1) \tag{6.6}$$

where $s(r_{ij})$ is the social force strength defined in Eq. 6.2 and $\mathbf{d}_{ij} = \left(\mathbf{l}_j^k - \mathbf{l}_i^k\right)/r_{ij}$ is the unit vector pointing from $\mathbf{l}_i^k$ to $\mathbf{l}_j^k$. Besides, $rand(1,-1)$ is a randomly generated number between 1 and -1. Likewise, $\rho\left(\mathbf{l}_i^k, \mathbf{l}_j^k\right)$ is the dominance function that calculates the dominance value of the most dominant individual from $\mathbf{l}_j^k$ and $\mathbf{l}_i^k$. In order to operate $\rho(\mathbf{l}_i^k, \mathbf{l}_j^k)$, all the individuals from $\mathbf{L}^k\left(\{\mathbf{l}_1^k, \mathbf{l}_2^k, \ldots, \mathbf{l}_N^k\}\right)$ are ranked according to their fitness values. The ranks are assigned so that the best individual receives the rank 0 (zero) whereas the worst individual obtains the rank $N - 1$. Therefore, the function $\rho(\mathbf{l}_i^k, \mathbf{l}_j^k)$ is defined as follows:

$$\rho(\mathbf{l}_i^k, \mathbf{l}_j^k) = \begin{cases} e^{-\left(5\cdot \text{rank}(\mathbf{l}_i^k)/N\right)} & \text{if } \text{rank}(\mathbf{l}_i^k) < \text{rank}(\mathbf{l}_j^k) \\ e^{-\left(5\cdot \text{rank}(\mathbf{l}_j^k)/N\right)} & \text{if } \text{rank}(\mathbf{l}_i^k) > \text{rank}(\mathbf{l}_j^k) \end{cases} \tag{6.7}$$

where the function rank(α) delivers the rank of the α-individual. According to Eq. 6.7, $\rho(\mathbf{l}_i^k, \mathbf{l}_j^k)$ yields a value within the interval (1, 0). Its maximum value of one in $\rho(\mathbf{l}_i^k, \mathbf{l}_j^k)$ is reached when either individual $\mathbf{l}_j^k$ or $\mathbf{l}_i^k$ is the best element of the population $\mathbf{L}^k$ regarding their fitness values. On the other hand, a value close to zero is obtained when both individuals $\mathbf{l}_j^k$ and $\mathbf{l}_i^k$ possess quite bad fitness values. Figure 6.3 shows the behavior of $\rho(\mathbf{l}_i^k, \mathbf{l}_j^k)$ considering 100 individuals. In the Figure, it is assumed that $\mathbf{l}_i^k$ represents one of the 99 individuals with ranks between 0 and 98 whereas $\mathbf{l}_j^k$ is fixed to the element with the worst fitness value (rank 99).

Under the incorporation of $\rho(\mathbf{l}_i^k, \mathbf{l}_j^k)$ in Eq. 6.6, social forces are magnified or weakened depending on the best fitness value (the most dominant) of the individuals involved in the repulsion-attraction process.

Finally, the total social force on each individual $\mathbf{l}_i^k$ is modeled as the superposition of all of the pairwise interactions exerted over it:

$$\mathbf{S}_i^m = \sum_{\substack{j=1 \\ j \neq i}}^{N} \mathbf{s}_{ij}^m \tag{6.8}$$

Therefore, the change of position $\Delta\mathbf{l}_i$ is considered as the total social force experimented by $\mathbf{l}_i^k$ as the superposition of all of the pairwise interactions. Thus, $\Delta\mathbf{l}_i$ is defined as follows:

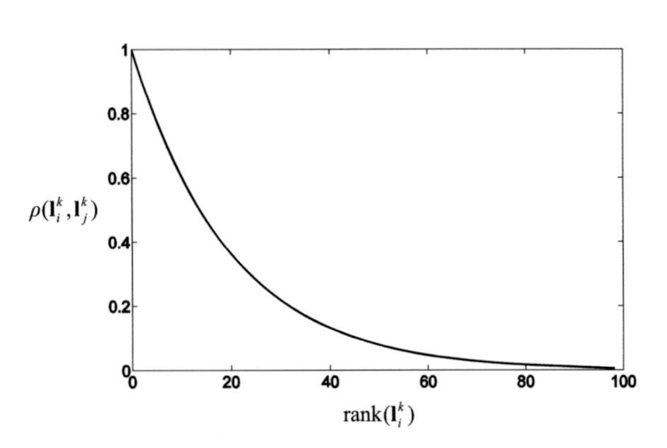

Fig. 6.3 Behavior of $\rho(\mathbf{l}_i^k, \mathbf{l}_j^k)$ considering 100 individuals

$$\Delta \mathbf{l}_i = \mathbf{S}_i^m \tag{6.9}$$

After calculating the new positions $\mathbf{P}(\{\mathbf{p}_1, \mathbf{p}_2, \ldots, \mathbf{p}_N\})$ of the population $\mathbf{L}^k(\{\mathbf{l}_1^k, \mathbf{l}_2^k, \ldots, \mathbf{l}_N^k\})$, the final positions $\mathbf{F}(\{\mathbf{f}_1, \mathbf{f}_2, \ldots, \mathbf{f}_N\})$ must be calculated. The idea is to admit only the changes that guarantee an improvement in the search strategy. If the fitness value of $\mathbf{p}_i(f(\mathbf{p}_i))$ is better than $\mathbf{l}_i^k(f(\mathbf{l}_i^k))$, then $\mathbf{p}_i$ is accepted as the final solution. Otherwise, $\mathbf{l}_i^k$ is retained. This procedure can be resumed by the following statement (considering a minimization problem):

$$\mathbf{f}_i = \begin{cases} \mathbf{p}_i & \text{if } f(\mathbf{p}_i) < f(\mathbf{l}_i^k) \\ \mathbf{l}_i^k & \text{otherwise} \end{cases} \tag{6.10}$$

In order to illustrate the performance of the solitary operator, Fig. 6.4 presents a simple example with the solitary operator being iteratively applied. It is assumed a population of 50 different members (N = 50) which adopt a concentrated configuration as initial condition (Fig. 6.4a). As a consequence of the social forces, the set

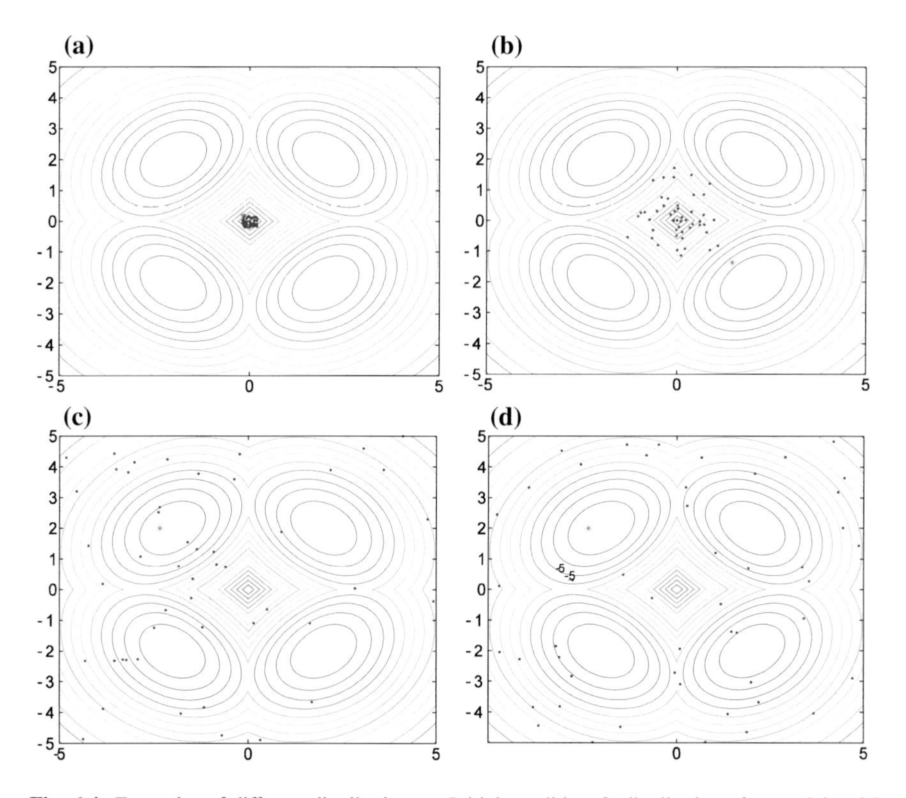

Fig. 6.4 Examples of different distributions. **a** Initial condition, **b** distribution after applying 25, **c** 50 and **d** 100 operations

of elements tends to distribute throughout the search space. Examples of different distributions are shown in Fig. 6.4b–d after applying 25, 50 and 100 different solitary operations, respectively.

6.3.2 Social Operation (B)

The social procedure represents the exploitation phase of the LS algorithm. Exploitation is the process of refining existent individuals within a small neighborhood in order to improve their solution quality.

The social procedure is a selective operation which is applied only to a subset $\mathbf{E}$ of the final positions F (where $\mathbf{E} \subseteq \mathbf{F}$). The operation starts by sorting F with respect to fitness values, storing the sorted elements in a temporal population $\mathbf{B} = \{\mathbf{b}_1, \mathbf{b}_2, \ldots, \mathbf{b}_N\}$. The elements in B are sorted so that the best individual receives the position $\mathbf{b}_1$ whereas the worst individual obtains the location $\mathbf{b}_N$. Therefore, the subset $\mathbf{E}$ is integrated by only the first g locations of B (promising solutions). Under this operation, a subspace C_j is created around each selected particle $\mathbf{f}_j \in \mathbf{E}$. The size of C_j depends on the distance e_d which is defined as follows:

$$e_d = \frac{\sum_{q=1}^{n} \left(ub_q - lb_q\right)}{n} \cdot \beta \tag{6.11}$$

where ub_q and lb_q are the upper and lower bounds in the q-th dimension, n is the number of dimensions of the optimization problem, whereas $\beta \in [0, 1]$ is a tuning factor. Therefore, the limits of C_j can be modeled as follows:

$$\begin{aligned} uss_j^q &= b_{j,q} + e_d \\ lss_j^q &= b_{j,q} - e_d \end{aligned} \tag{6.12}$$

where uss_j^q and lss_j^q are the upper and lower bounds of the q-th dimension for the subspace C_j, respectively.

Considering the subspace C_j around each element $\mathbf{f}_j \in \mathbf{E}$, a set of h new particles $\left(\mathbf{M}_j^h = \left\{\mathbf{m}_j^1, \mathbf{m}_j^2, \ldots, \mathbf{m}_j^h\right\}\right)$ are randomly generated inside bounds fixed by Eq. 6.12. Once the h samples are generated, the individual $\mathbf{l}_j^{k+1}$ of the next population $\mathbf{L}^{k+1}$ must be created. In order to calculate $\mathbf{l}_j^{k+1}$, the best particle $\mathbf{m}_j^{best}$, in terms of its fitness value from the h samples $\left(\text{where } \mathbf{m}_j^{best} \in \left[\mathbf{m}_j^1, \mathbf{m}_j^2, \ldots, \mathbf{m}_j^h\right]\right)$, is compared to $\mathbf{f}_j$. If $\mathbf{m}_j^{best}$ is better than $\mathbf{f}_j$ according to their fitness values, $\mathbf{l}_j^{k+1}$ is updated with $\mathbf{m}_j^{best}$, otherwise $\mathbf{f}_j$ is selected. The elements of F that have not been processed by the procedure ($\mathbf{f}_w \notin \mathbf{E}$) transfer their corresponding values to $\mathbf{L}^{k+1}$ with no change.

The social operation is used to exploit only prominent solutions. According to the presented method, inside each subspace C_j, h random samples are selected. Since the number of selected samples at each subspace is very small (typically $h < 4$), the use of this operator substantially reduces the number of fitness function evaluations.

In order to demonstrate the social operation, a numerical example has been set by applying the presented process to a simple function. Such function considers the interval of $-3 \leq d_1, d_2 \leq 3$ whereas the function possesses one global maxima of value 6.1 at $(0, 1.6)$. Notice that d_1 and d_2 correspond to the axis coordinates (commonly x and y). For this example, it is assumed a final position population F of six 2-dimensional members (N = 6). Figure 6.5 shows the initial configuration of the presented example, with the black points representing half of the particles with the best fitness values (the first three element of B, g = 3) whereas the grey points ($\mathbf{f}_2, \mathbf{f}_4, \mathbf{f}_6 \notin \mathbf{E}$) correspond to the remaining individuals. From Fig. 6.5, it can be seen that the social procedure is applied to all black particles ($\mathbf{f}_5 = \mathbf{b}_1$, $\mathbf{f}_3 = \mathbf{b}_2$ and $\mathbf{f}_1 = \mathbf{b}_3$, $\mathbf{f}_5, \mathbf{f}_3, \mathbf{f}_1 \in \mathbf{E}$) yielding two new random particles (h = 2), which are characterized by white points $\mathbf{m}_1^1$, $\mathbf{m}_1^2$, $\mathbf{m}_3^1$, $\mathbf{m}_3^2$, $\mathbf{m}_5^1$ and $\mathbf{m}_5^2$, for each black point inside of their corresponding subspaces C_1, C_3 and C_5. Considering the particle $\mathbf{f}_3$ in Fig. 6.5, the particle $\mathbf{m}_3^2$ corresponds to the best particle ($\mathbf{m}_3^{best}$) from the two randomly generated particles (according to their fitness values) within C_3. Therefore, the particle $\mathbf{m}_3^{best}$ will substitute $\mathbf{f}_3$ in the individual $\mathbf{l}_3^{k+1}$ for the next generation, since it holds a better fitness value than $\mathbf{f}_3$ $\left(f(\mathbf{f}_3) < f(\mathbf{m}_3^{best})\right)$.

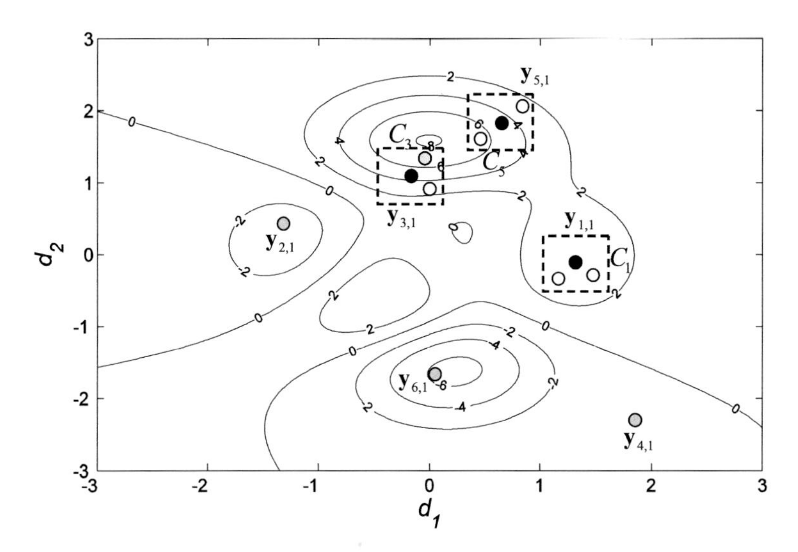

Fig. 6.5 Operation of the social procedure

6.3.3 Complete LS Algorithm

LS is a simple algorithm with only five adjustable parameters: the strength of attraction F, the attractive length L, number of promising solutions g, the population size N, and the number of generations gen. The operation of LS is divided into three parts: Initialization of the solitary and social operations. In the initialization ($k = 0$), the first population $\mathbf{L}^0$ $(\{\mathbf{l}_1^0, \mathbf{l}_2^0, \ldots, \mathbf{l}_N^0\})$ is produced. The values $\left\{l_{i,1}^0, l_{i,2}^0, \ldots, l_{i,n}^0\right\}$ of each individual $\mathbf{l}_i^k$ and each dimension d are randomly and uniformly distributed between the pre-specified lower initial parameter bound lb_d and the upper initial parameter bound ub_d.

$$\begin{aligned} & l_{i,j}^0 = lb_d + \text{rand} \cdot (ub_d - lb_d); \\ & i = 1, 2, \ldots, N; \quad d = 1, 2, \ldots, n. \end{aligned} \tag{6.13}$$

In the evolution process, the solitary (A) and social (B) operations are iteratively applied until the number of iterations $k = gen$ has been reached. The complete LS procedure is illustrated in the Algorithm 6.1.

Algorithm 6.1. Locust Search (LS) algorithm		
1:	Input: F, L, g, N and gen	
2:	Initialize $\mathbf{L}^0$ (k=0)	
3:	until ($k = gen$)	
5:	$\mathbf{F} \leftarrow$ SolitaryOperation($\mathbf{L}^k$)	Solitary operator (3.1)
6:	$\mathbf{L}^{k+1} \leftarrow$ Social Operation($\mathbf{L}^k$, $\mathbf{F}$)	Social operator (3.2)
8:	k=k+1	
7:	end until	

6.3.4 Discussion About the LS Algorithm

Evolutionary algorithms (EA) have been widely employed for solving complex optimization problems. These methods are found to be more powerful than conventional methods that are based on formal logics or mathematical programming [30]. In the EA algorithm, search agents have to decide whether to explore unknown search positions or to exploit already tested positions in order to improve their solution quality. Pure exploration degrades the precision of the evolutionary process but increases its capacity to find new potentially solutions. On the other hand, pure exploitation allows refining existent solutions but adversely drives the process to local optimal solutions. Therefore, the ability of an EA to find a global optimal solution depends on its capacity to find a good balance between the exploitation of found-so-far elements and the exploration of the search space [31]. So far, the exploration-exploitation dilemma has been an unsolved issue within the framework of evolutionary algorithms.

Most of swarm algorithms and other evolutionary algorithms tend to exclusively concentrate the individuals in the current best positions. Under such circumstances, such algorithms seriously limit their exploration-exploitation capacities.

Different to most of existent evolutionary algorithms, in the presented approach, the modeled behavior explicitly avoids the concentration of individuals in the current best positions. Such fact allows not only to emulate the cooperative behavior of the locust colony in a good realistic way, but also to incorporate a computational mechanism to avoid critical flaws that are commonly present in the popular PSO and DE algorithms, such as the premature convergence and the incorrect exploration-exploitation balance.

6.4 Experimental Results

A comprehensive set of 13 functions, collected from Refs. [32–37], has been used to test the performance of the presented approach. Tables 6.1 and 6.2 present the benchmark functions used in our experimental study. Such functions are classified into two different categories: unimodal test functions (Table 6.1) and multimodal test functions (Table 6.2). In these tables, n is the function dimension, f_{opt} is the minimum value of the function, with S being a subset of R^n. The optimum location ($\mathbf{x}_{opt}$) for functions in Tables 6.1 and 6.2, are in $[0]^n$, except for f_5, f_{12}, f_{13} with $\mathbf{x}_{opt}$ in $[1]^n$ and f_8 in $[420.96]^n$. A detailed description of optimum locations is given in Table 6.4.

Table 6.1 Unimodal test functions

Test function	S	f_{opt}
$f_1(\mathbf{x}) = \sum_{i=1}^{n} x_i^2$	$[-100, 100]^n$	0
$f_2(\mathbf{x}) = \sum_{i=1}^{n} \lvert x_i \rvert + \prod_{i=1}^{n} \lvert x_i \rvert$	$[-10, 10]^n$	0
$f_3(\mathbf{x}) = \sum_{i=1}^{n} \left(\sum_{j=1}^{i} x_j \right)^2$	$[-100, 100]^n$	0
$f_4(\mathbf{x}) = \max_i \{ \lvert x_i \rvert, 1 \le i \le n \}$	$[-100, 100]^n$	0
$f_5(\mathbf{x}) = \sum_{i=1}^{n-1} \left[100 (x_{i+1} - x_i^2)^2 + (x_i - 1)^2 \right]$	$[-30, 30]^n$	0
$f_6(\mathbf{x}) = \sum_{i=1}^{n} (x_i + 0.5)^2$	$[-100, 100]^n$	0
$f_7(\mathbf{x}) = \sum_{i=1}^{n} i x_i^4 + rand(0, 1)$	$[-1.28, 1.28]^n$	0

Table 6.2 Multimodal test functions

Test function	S	f_{opt}
$f_8(\mathbf{x}) = \sum_{i=1}^{n} -x_i \sin\left(\sqrt{\lvert x_i \rvert}\right)$	$[-500, 500]^n$	$-418.98 * n$
$f_9(\mathbf{x}) = \sum_{i=1}^{n} \left[x_i^2 - 10\cos(2\pi x_i) + 10\right]$	$[-5.12, 5.12]^n$	0
$f_{10}(\mathbf{x}) = -20\exp\left(-0.2\sqrt{\frac{1}{n}\sum_{i=1}^{n} x_i^2}\right) - \exp\left(\frac{1}{n}\sum_{i=1}^{n}\cos(2\pi x_i)\right) + 20$	$[-32, 32]^n$	0
$f_{11}(\mathbf{x}) = \frac{1}{4000}\sum_{i=1}^{n} x_i^2 - \prod_{i=1}^{n}\cos\left(\frac{x_i}{\sqrt{i}}\right) + 1$	$[-600, 600]^n$	0
$f_{12}(\mathbf{x}) = \frac{\pi}{n}\left\{10\sin(\pi y_1) + \sum_{i=1}^{n-1}(y_i - 1)^2\left[1 + 10\sin^2(\pi y_{i+1})\right] + (y_n - 1)^2\right\}\cdots$ $+ \sum_{i=1}^{n} u(x_i, 10, 100, 4)$ $y_i = 1 + \frac{x_i + 1}{4} \quad u(x_i, a, k, m) = \begin{cases} k(x_i - a)^m & x_i > a \\ 0 & -a < x_i < a \\ k(-x_i - a)^m & x_i < -a \end{cases}$	$[-50, 50]^n$	0
$f_{13}(\mathbf{x}) = 0.1\left\{\sin^2(3\pi x_1) + \sum_{i=1}^{n}(x_i - 1)^2\left[1 + \sin^2(3\pi x_i + 1)\right]\cdots\right.$ $\left. + (x_n - 1)^2\left[1 + \sin^2(2\pi x_n)\right]\right\} + \sum_{i=1}^{n} u(x_i, 5, 100, 4)$	$[-50, 50]^n$	0

In Table 6.1, n is the dimension of function, f_{opt} is the minimum value of the function, and S is a subset of R^n. The optimum location ($\mathbf{x}_{opt}$) for functions in Table 6.1 is in $[0]^n$, except for f_5 with $\mathbf{x}_{opt}$ in $[1]^n$.

The optimum location ($\mathbf{x}_{opt}$) for functions in Table 6.2, are in $[0]^n$, except for f_8 in $[420.96]^n$ and $f_{12} - f_{13}$ in $[1]^n$.

6.4.1 Performance Comparison

We have applied the LS algorithm to 13 functions whose results have been compared to those produced by the Particle Swarm Optimization (PSO) method [3] and the Differential Evolution (DE) algorithm [11]. These are considered as the most popular algorithms for many optimization applications. In all comparisons, the population has been set to 40 ($N = 40$) individuals. The maximum iteration number for all functions has been set to 1000. Such stop criterion has been selected to maintain compatibility to similar works reported in the literature [34, 35].

The parameter setting for each of the algorithms in the comparison is described as follows:

1. PSO: In the algorithm, $c_1 = c_2 = 2$ while the inertia factor (ω) is decreased linearly from 0.9 to 0.2.
2. DE: The DE/Rand/1 scheme is employed. The parameter settings follow the instructions in [11]. The crossover probability is $CR = 0.9$ and the weighting factor is $F = 0.8$.
3. In LS, F and L are set to 0.6 and L, respectively. Besides, g is fixed to 20 ($N/2$) whereas gen and N are configured to 1000 and 40, respectively. Once such parameters have been experimentally determined, they are kept for all experiments in this section.

6.4.1.1 Unimodal Test Functions

A non-parametric statistical significance proof known as the Wilcoxon's rank sum test for independent samples [38, 39] has been conducted with an 5 % significance level, over the "average best-so-far" data of Table 6.3. Table 6.4 reports the p-values produced by Wilcoxon's test for the pair-wise comparison of the "average best so-far" of two groups. Such groups are formed by LS vs. PSO and LS vs. DE. As a null hypothesis, it is assumed that there is no significant difference between mean values of the two algorithms. The alternative hypothesis considers a significant difference between the "average best-so-far" values of both approaches. All p-values reported in the table are less than 0.05 (5 % significance level) which is a strong evidence against the null hypothesis, indicating that the LS results are

Table 6.3 Minimization results from the benchmark functions test in Table 8.1 with $n = 30$

		PSO	DE	LS
f_1	ABS	1.66×10^{-1}	6.27×10^{-3}	4.55×10^{-4}
	MBS	0.23	5.85×10^{-3}	2.02×10^{-4}
	SD	3.79×10^{-1}	1.68×10^{-1}	6.98×10^{-4}
f_2	ABS	4.83×10^{-1}	2.02×10^{-1}	5.41×10^{-3}
	MBS	0.53	1.96×10^{-1}	5.15×10^{-3}
	SD	1.59×10^{-1}	0.66	1.45×10^{-2}
f_3	ABS	2.75	5.72×10^{-1}	1.61×10^{-3}
	MBS	3.16	6.38×10^{-1}	1.81×10^{-3}
	SD	1.01	0.15	1.32×10^{-3}
f_4	ABS	1.84	0.11	1.05×10^{-2}
	MBS	1.79	0.10	1.15×10^{-2}
	SD	0.87	0.05	6.63×10^{-3}
f_5	ABS	3.07	2.39	4.11×10^{-2}
	MBS	3.03	2.32	3.65×10^{-2}
	SD	0.42	0.36	2.74×10^{-3}
f_6	ABS	6.36	6.51	5.88×10^{-2}
	MBS	6.19	6.60	5.17×10^{-2}
	SD	0.74	0.87	1.67×10^{-2}
f_7	ABS	6.14	0.12	2.71×10^{-2}
	MBS	2.76	0.14	1.10×10^{-2}
	SD	0.73	0.02	1.18×10^{-2}

Maximum number of iterations = 1000

Table 6.4 p-values produced by Wilcoxon's test that compares LS versus PSO and DE over the "average best-so-far" values from Table 6.1

LS versus	PSO	DE
f_1	1.83×10^{-4}	1.73×10^{-2}
f_2	3.85×10^{-3}	1.83×10^{-4}
f_3	1.73×10^{-4}	6.23×10^{-3}
f_4	2.57×10^{-4}	5.21×10^{-3}
f_5	4.73×10^{-4}	1.83×10^{-3}
f_6	6.39×10^{-5}	2.15×10^{-3}
f_7	1.83×10^{-4}	2.21×10^{-3}

statistically significant and that it has not occurred by coincidence (i.e. due to the normal noise contained in the process).

6.4.1.2 Multimodal Test Functions

Multimodal functions possess many local minima which makes the optimization a difficult task to be accomplished. For multimodal functions, the final results are more important since they reflect the algorithm's ability to escape from poor local optima and locate a near-global optimum. The algorithms have been applied over functions f_8 to f_{13} where the number of local minima increases exponentially as the dimension of the function increases. The dimension of such functions is set to 30. The results are averaged over 30 runs, reporting the performance indexes in Table 6.5 as it follows: the average best-so-far solution (ABS), the median of the best solution in the last iteration (MBS) and the standard deviation (SD). Likewise, p-values of the Wilcoxon signed-rank test of 30 independent runs are listed in Table 6.6.

For f_9, f_{10}, f_{11} and f_{12}, LS yields a much better solution than others algorithms. However, for functions f_8 and f_{13}, LS produces similar results to DE. The Wilcoxon rank test results, that are presented in Table 6.6, show that LS performed better than

Table 6.5 Minimization results from the benchmark functions test in Table 6.2 with $n = 30$

		PSO	DE	LS
f_8	ABS	-6.7×10^3	-1.26×10^4	-1.26×10^4
	MBS	-5.4×10^3	-1.24×10^4	-1.23×10^4
	SD	6.3×10^2	3.7×10^2	1.1×10^2
f_9	ABS	14.8	4.01×10^{-1}	2.49×10^{-3}
	MBS	13.7	2.33×10^{-1}	3.45×10^{-3}
	SD	1.39	5.1×10^{-2}	4.8×10^{-4}
f_{10}	ABS	14.7	4.66×10^{-2}	2.15×10^{-3}
	MBS	18.3	4.69×10^{-2}	1.33×10^{-3}
	SD	1.44	1.27×10^{-2}	3.18×10^{-4}
f_{11}	ABS	12.01	1.15	1.47×10^{-4}
	MBS	12.32	0.93	3.75×10^{-4}
	SD	3.12	0.06	1.48×10^{-5}
f_{12}	ABS	6.87×10^{-1}	3.74×10^{-1}	5.58×10^{-3}
	MBS	4.66×10^{-1}	3.45×10^{-1}	5.10×10^{-3}
	SD	7.07×10^{-1}	1.55×10^{-1}	4.18×10^{-4}
f_{13}	ABS	1.87×10^{-1}	1.81×10^{-2}	1.78×10^{-2}
	MBS	1.30×10^{-1}	1.91×10^{-2}	1.75×10^{-2}
	SD	5.74×10^{-1}	1.66×10^{-2}	1.64×10^{-3}

Maximum number of iterations = 1000

Table 6.6 p-values produced by Wilcoxon's test comparing LS vs PSO and DE over the "average best-so-far" values from Table 6.5

LS vs	PSO	DE
f_8	1.83×10^{-4}	0.061
f_9	1.17×10^{-4}	2.41×10^{-4}
f_{10}	1.43×10^{-4}	3.12×10^{-3}
f_{11}	6.25×10^{-4}	1.14×10^{-3}
f_{12}	2.34×10^{-5}	7.15×10^{-4}
f_{13}	4.73×10^{-4}	0.071

PSO and DE considering four problems $f_9 - f_{12}$, whereas, from a statistical viewpoint, there is not difference between results from LS and DE for f_8 and f_{13}.

6.5 Summary

In this chapter, a novel swarm algorithm, called the Locust Search (LS) has been presented for solving optimization tasks. The LS algorithm is based on the behavioral simulation of swarms of locusts. In the presented algorithm, individuals emulate a group of locusts which interact to each other based on the biological laws of the cooperative swarm. The algorithm considers two different behaviors: solitary and social. Depending on the selected behavioral scheme, each individual is conducted by a set of evolutionary operators which mimic the different cooperative conducts that are typically found in the swarm.

Different to most of existent swarm algorithms, the behavioral model in the presented approach explicitly avoids the concentration of individuals in the current best positions. Such fact allows not only to emulate in a good realistic way the cooperative behavior of the locust colony, but also to incorporate a computational mechanism to avoid critical flaws that are commonly present in the popular PSO and DE algorithms, such as the premature convergence and the incorrect exploration-exploitation balance.

LS has been experimentally tested considering a suite of 13 benchmark functions. The performance of LS has been compared to the following algorithms: the Particle Swarm Optimization method (PSO) [3], and the Differential Evolution (DE) algorithm [11]. Results have produced an acceptable performance of the presented method in terms of the solution quality for all tested benchmark functions.

The LS remarkable performance is associated with two different reasons: (i) the solitary operator allows a better particle distribution in the search space, increasing the algorithm's ability to find the global optima; and (ii) the use of the social operation provides of a simple exploitation operator that intensifies the capacity of finding better solutions during the evolution process.

References

1. Bonabeau, E., Dorigo, M., Theraulaz, G.: Swarm Intelligence: From Natural to Artificial Systems. Oxford University Press Inc, New York, NY, USA (1999)
2. Kassabalidis, I., El-Sharkawi, M.A., II Marks, R.J., Arabshahi, P., Gray, A.A.: Swarm intelligence for routing in communication networks. Global Telecommunications Conference, GLOBECOM'01, IEEE, 2001, vol. 6, pp. 3613–3617
3. Kennedy, J., Eberhart, R.: Particle swarm optimization. In: Proceedings of the 1995 IEEE International Conference on Neural Networks, vol. 4, pp. 1942–1948, Dec 1995

4. Karaboga, D.: An Idea Based on Honey Bee Swarm for Numerical Optimization. TechnicalReport-TR06. Engineering Faculty, Computer Engineering Department, Erciyes University (2005)
5. Passino, K.M.: Biomimicry of bacterial foraging for distributed optimization and control. IEEE Cont. Syst. Mag. **22**(3), 52–67 (2002)
6. Hossein, A., Hossein-Alavi, A.: Krill herd: a new bio-inspired optimization algorithm. Commun. Nonlinear Sci. Numer. Simulat. **17**, 4831–4845 (2012)
7. Yang, X.S: Engineering optimization: an introduction with metaheuristic applications. Wiley (2010)
8. Yang, X.S., Deb, S.: Proceedings of World Congress on Nature & Biologically Inspired Computed, pp. 210–214. IEEE Publications, India (2009)
9. Cuevas, E., Cienfuegos, M., Zaldívar, D., Pérez-Cisneros, M.: A swarm optimization algorithm inspired in the behavior of the social-spider. Expert Syst. Appl. **40**(16), 6374–6384 (2013)
10. Cuevas, E., González, M., Zaldivar, D., Pérez-Cisneros, M., García, G.: An algorithm for global optimization inspired by collective animal behaviour. Discrete Dyn. Nat. Soc., art. no. 638275 (2012)
11. Storn, R., Price, K.: Differential Evolution—a simple and efficient adaptive scheme for global optimisation over continuous spaces. TechnicalReportTR-95–012, ICSI, Berkeley, CA (1995)
12. Bonabeau, E.: Social insect colonies as complex adaptive systems. Ecosystems **1**, 437–443 (1998)
13. Wang, Y., Li, B., Weise, T., Wang, J., Yuan, B., Tian, Q.: Self-adaptive learning based particle swarm optimization. Inf. Sci. **181**(20), 4515–4538 (2011)
14. Tvrdík, J.: Adaptation in differential evolution: a numerical comparison. Appl. Soft Comput. **9** (3), 1149–1155 (2009)
15. Wang, H., Sun, H., Li, C., Rahnamayan, S., Jeng-shyang, P.: Diversity enhanced particle swarm optimization with neighborhood. Inf. Sci. **223**, 119–135 (2013)
16. Gong, W., Fialho, Á., Cai, Z., Li, H.: Adaptive strategy selection in differential evolution for numerical optimization: an empirical study. Inf. Sci. **181**(24), 5364–5386 (2011)
17. Gordon, D.: The Organization of work in social insect colonies. Complexity **8**(1), 43–46 (2003)
18. Kizaki, S., Katori, M.: A stochastic lattice model for locust outbreak. Phys. A **266**, 339–342 (1999)
19. Rogers, S.M., Cullen, D.A., Anstey, M.L., Burrows, M., Dodgson, T., Matheson, T., Ott, S.R., Stettin, K., Sword, G.A., Despland, E., Simpson, S.J.: Rapid behavioural gregarization in the desert locust, Schistocerca gregaria entails synchronous changes in both activity and attraction to conspecifics. J. Insect Physiol. **65**, 9–26 (2014)
20. Topaz, C.M., Bernoff, A.J., Logan, S., Toolson, W.: A model for rolling swarms of locusts. Eur. Phys. J. Spec. Top. **157**, 93–109 (2008)
21. Topaz, C.M., D'Orsogna, M.R., Edelstein-Keshet, L., Bernoff, A.J.: Locust dynamics behavioral phase change and swarming. PLoS Comput. Biol. **8**(8), 1–11 (2012)
22. Oster, G., Wilson, E.: Caste and ecology in the social insects. Princeton University Press, Princeton, N.J. (1978)
23. Hölldobler, B., Wilson, E.O.: Journey to the Ants: A Story of Scientific Exploration. ISBN 0-674-48525-4 (1994)
24. Hölldobler, B.,Wilson, E.O.: The Ants. Harvard University Press. ISBN 0-674-04075-9 (1990)
25. Tanaka, S., Nishide, Y.: Behavioral phase shift in nymphs of the desert locust, Schistocerca gregaria: special attention to attraction/avoidance behaviors and the role of serotonin. J. Insect Physiol. **59**, 101–112 (2013)
26. Gaten, Edward, Huston, Stephen J., Dowse, Harold B., Matheson, Tom: Solitary and gregarious locusts differ in circadian rhythmicity of a visual output neuron. J. Biol. Rhythms **27**(3), 196–205 (2012)

27. Benaragama, I., Gray, J.R.: Responses of a pair of flying locusts to lateral looming visual stimuli. J. Comp. Physiol. A. **200**(8), 723–738 (2014)
28. Sergeev, MG.: Distribution Patterns of Grasshoppers and Their Kin in the Boreal Zone, vol. 2011, 9 p. Article ID 324130 (2011)
29. Ely, SO., Njagi, PGN., Bashir, MO., El-Amin, SET., Hassanali, A.: Diel behavioral activity patterns in adult solitarious desert locust, Schistocerca gregaria (Forskål). Psyche **2011**, p. 9 (2011) Article ID 459315
30. Yang, X.-S.: Nature-inspired metaheuristic algorithms. Luniver Press, Beckington (2008)
31. Cuevas, E., Echavarría, A., Ramírez-Ortegón, M.A.: An optimization algorithm inspired by the States of Matter that improves the balance between exploration and exploitation. Appl. Intell. **40**(2), 256–272 (2014)
32. Ali, M.M., Khompatraporn, C., Zabinsky, Z.B.: A numerical evaluation of several stochastic algorithms on selected continuous global optimization test problems. J. Glob. Optim. **31**(4), 635–672 (2005)
33. Chelouah, R., Siarry, P.: A continuous genetic algorithm designed for the global optimization of multimodal functions. J. Heuristics **6**(2), 191–213 (2000)
34. Herrera, F., Lozano, M., Sánchez, A.M.: A taxonomy for the crossover operator for real-coded genetic algorithms: an experimental study. Int. J. Intell. Syst. **18**(3), 309–338 (2003)
35. Laguna, M., Martí, R.: Experimental testing of advanced scatter search designs for global optimization of multimodal functions. J. Glob. Optim. **33**(2), 235–255 (2005)
36. Lozano, M., Herrera, F., Krasnogor, N., Molina, D.: Real-coded memetic algorithms with crossover hill-climbing. Evol. Comput. **12**(3), 273–302 (2004)
37. Moré, J.J., Garbow, B.S., Hillstrom, K.E.: Testing unconstrained optimization software. ACM Trans. Math. Softw. **7**(1), 17–41 (1981)
38. Wilcoxon, F.: Individual comparisons by ranking methods. Biometrics **1**, 80–83 (1945)
39. Garcia, S., Molina, D., Lozano, M., Herrera, F.: A study on the use of non-parametric tests for analyzing the evolutionary algorithms' behaviour: a case study on the CEC'2005 special session on real parameter optimization. J Heurist (2008). doi:10.1007/s10732-008-9080-4

Chapter 7
Reduction of Function Evaluations by using an evolutionary computation algorithm

7.1 Introduction

Evolutionary algorithms (EAs) are optimization tools that have demonstrated their efficiency and flexibility in many application domains. They can be applied to a variety of problems to which traditional optimization methods do not apply. Some examples of such approaches include Genetic algorithms (RGA) [1], Particle Swarm Optimization (PSO) [2], Clonal selection Algorithm (CSA) [3] and Evolutionary strategies (ES) [4]. However, it is well known that the main drawback of EAs is their high computational cost in terms of number of evaluations of the fitness function [5]. This problem becomes even more critical when the fitness evaluation is computationally expensive [6].

The problem of reducing the number of function calculations has already been faced in the field of EA and is better known as evolution control or fitness estimation [7]. Under these approaches, the idea is to replace the costly objective function by an approximated model which is easier to calculate. In addition to the approximated model, it is incorporated a fitness estimation strategy that decides the amount of individuals to be evaluated with the original fitness function and the amount of individuals to be calculated with the approximated model. In the practice, it is very difficult to construct an approximated model which can be globally deal with the high dimensionality, ill distribution and limited number of training samples. Several approaches have been proposed in the literature where popular EAs are combined with approximated models. Some examples include the integration of CSA with Nearest neighbors (CSA-NN) [8], PSO with radial basis functions (PSO-RBF) [9] and ES with kriging models (ES-K) [10]. Although approximated models have exhibited relative effectiveness, experimental studies [11] have demonstrated that their use degrades the search effectiveness of the original EAs, producing frequently inaccurate solutions [12]. Since no single approximated model and evolutionary method is optimal for all problems, so far, this dilemma has been an unsolved issue within the framework of evolutionary algorithms.

E. Cuevas et al., *Advances of Evolutionary Computation: Methods and Operators*, Studies in Computational Intelligence 629,
DOI 10.1007/978-3-319-28503-0_7

In an EA, increasing the population size positively influences the quality of the produced individuals, but adversely affects its computational cost [13]. Traditionally, population size is set in advance to a specified value and remains fixed through the entire execution of the algorithm. If the population size is too small the EA may converge too quickly affecting severely the solution quality [14]. On the other hand, if it is too large the EA may present a prohibitive computational cost [13]. Therefore, an appropriate population size allows maintaining a balance between computational cost and effectiveness of the algorithm. In order to solve such a problem, several approaches have been proposed for adapting dynamically the population size. These methods are grouped into three categories [15]: (i) methods that increment or decrement the number of individuals according to a fixed criterion; (ii) methods in which the number of individuals is modified according to the performance of the average fitness value and (iii) algorithms based on the population diversity.

To use either an alternative model or an adaptive population size approach is necessary but not sufficient to tackle the problem of reducing the number of function evaluations. Using a fitness estimation strategy during the evolution process without adapting the population size to improve the population diversity, makes the algorithm defenseless against the convergence to a false minimum and may result in poor exploratory characteristics of the algorithm [12]. On the other hand, adapting the population size without using a fitness estimation strategy leads to increase the computational cost [14]. Therefore, it does seem reasonable to incorporate both approaches into a single algorithm.

Recently, a new evolutionary method called the Adaptive Population with Reduced Evaluations (APRE) has been presented for solving image processing problems which are characterized by demanding an excessive number of function evaluations. APRE reduces the number of function evaluations through the use of two mechanisms: (1) adapting dynamically the size of the population and (2) incorporating a fitness estimation strategy that decides the amount of individuals to be evaluated with the original fitness function and the amount of individuals to be estimated by a very simple approximated model. APRE begins with an initial random population which will be used as a memory during the evolution process. After initialization, it is selected the elements to be evolved. Its number is automatically modified in each iteration. With the selected elements, a set of new individuals is generated as a consequence of the execution of the seeking operation. Afterwards, the memory is updated. For this process, the new individuals produced by the seeking operation compete against the memory elements to build the final memory configuration. Finally, a sample of the best elements contained in the final memory configuration is undergone to the refinement operation. This cycle is repeated until the maximum number the iterations has been reached.

Different to other approaches that use an already existent EA as framework, the APRE method has been completely designed to substantially reduce the computational cost without degrading its good search capacities. In this chapter, the performance of APRE as a global optimization algorithm is presented. In order to illustrate its proficiency and robustness, it is compared with other approaches that have been conceived to reduce the number of function evaluations. The comparison examines several standard benchmark functions which are commonly considered within the EA communities. Experimental results show that the APRE achieves the best balance in comparison to its counterparts with regard to the number of function evaluations and solution accuracy.

This chapter is organized as follows. In Sect. 7.2, the APRE algorithm and its characteristics are both described. Section 7.3 presents experimental results and a comparative study. Finally, in Sect. 7.4, some conclusions are discussed.

7.2 The Adaptive Population with Reduced Evaluations (APRE) Algorithm

APRE [16] is an evolutionary computation method, designed to reduce the number of function evolutions in image processing applications. The algorithm begins with an initial population which will be used as a memory during the evolution process. To each memory element, it is assigned a normalized fitness value called quality factor that indicates the solution capacity provided by the element.

In EAs, there exist two important properties that condition the search strategy: exploration and exploitation [17]. Exploration is the fact of visiting entirely new points of a search space, whilst exploitation is the process of refining those points of a search space within the neighborhood of previously visited locations in order to improve their solution quality. Pure exploration degrades the precision of the evolutionary process but increases its capacity to find new potentially solutions [18]. On the other hand, pure exploitation allows refining existent solutions but adversely drives the process to fall in local optimal solutions [19]. Therefore, the ability of an EA to find a global optimal solution depends on its capacity to find a good balance between the exploitation of found-so-far elements and the exploration of the search space [20]. As a search strategy, APRE implements two operations: “seeking” and “refinement” which try to positively influence the exploration and exploitation process, respectively.

APRE is an iterative process in which several actions are executed. First, it is computed the number of memory elements to be evolved. Such number is automatically modified in each iteration. Then, a set of new individuals is generated as a consequence of the execution of the seeking operation. Afterwards, the memory is updated. For this process, the new individuals produced by the seeking operation compete against the memory elements to build the final memory configuration. Finally, a sample of the best elements contained in the final memory configuration is undergone to the refinement operation.

The complete APRE process can be divided in four phases: Initialization, selecting the population to be evolved, seeking, fitness estimation strategy, memory updating and refinement. They are discussed in the following sections.

7.2.1 Initialization

The APRE algorithm begins by initializing (k = 0) a population M(k) of N_p elements ($\mathbf{M}(k) = \{\mathbf{m}_1, \mathbf{m}_2, \ldots, \mathbf{m}_{N_p}\}$). Each element $\mathbf{m}_i(m_{i,1}, m_{i,2}, \ldots m_{i,D})$ is a D-dimensional vector containing the parameter to be optimized. Its values are randomly and uniformly distributed between the pre-specified lower initial parameter bound p_j^{low} and the upper initial parameter bound p_j^{high}, just as it described by the following expression:

$$m_{i,j} = p_j^{low} + \text{rand}(0,1) \cdot (p_j^{high} - p_j^{low}) \\ j = 1, 2, \ldots, D, \quad i = 1, 2, \ldots, N_p \tag{7.1}$$

being j and i the parameter and element indexes respectively. Hence, $m_{i,j}$ is the jth parameter of the ith element.

Each element $\mathbf{m}_i$ has two associated characteristics: a fitness value $J(\mathbf{m}_i)$ and a quality factor $Q(\mathbf{m}_i)$. The fitness value $J(\mathbf{m}_i)$ assigned to each element $\mathbf{m}_i$ can be calculated by using the original objective function $f(\mathbf{m}_i)$ or only estimated by using the APRE fitness estimation strategy $F(\mathbf{m}_i)$. In addition to the fitness value, it is also assigned to $\mathbf{m}_i$ a normalized fitness value called quality factor $Q(\mathbf{m}_i)$ ($Q(\cdot) \in [0,1]$) which is computed as follows:

$$Q(\mathbf{m}_i) = \frac{J(\mathbf{m}_i) - worst_{\mathbf{M}}}{best_{\mathbf{M}} - worst_{\mathbf{M}}} \tag{7.2}$$

$\mathbf{m}_i$. The values $worst_{\mathbf{M}}$ and $best_{\mathbf{M}}$ are defined as follows (considering a maximization problem):

$$best_{\mathbf{M}} = \max_{k \in \{1,2,\ldots,N_p\}} (J(\mathbf{m}_k)) \quad \text{and} \quad worst_{\mathbf{M}} = \min_{k \in \{1,2,\ldots,N_p\}} (J(\mathbf{m}_k)) \tag{7.3}$$

Since the mechanism by which an EA accumulates information regarding the objective function is an exact evaluation of the quality of each potential solution, initially, all the elements of M(k) are evaluated without considering the fitness estimation strategy. This fact is only allowed at this initial stage.

7.2.2 Selecting the Population to Be Evolved

At each k iteration, it must be selected which and how many elements from M(k) will be considered to build the population $\mathbf{P}^k$ to be evolved. Such selected elements will be undergone by the seeking and refinement operators in order to generate a set of new individuals. Therefore, two things need to be defined: the number of elements N_e^k to be selected and the criterion of selection.

7.2.2.1 The Number of Elements N_e^k to Be Selected

One of the processes used by APRE for reducing the number of function evaluations is to modify dynamically the size of the population to be evolved. The objective is to operate with the optimal number of individuals that guarantee the correct efficiency of the algorithm. Therefore, the method implements an adaptation mechanism to vary the population size N_e^k depending on the performance of the optimization process. In an EA, increasing the population size positively influences the quality of the produced individuals, but adversely affects its computational cost [13]. Under such circumstances, the idea is to improve the algorithm's performance by increasing the population if the quality of the already produced individuals is "bad". On the contrary, if the quality of the already produced individuals is "good", the population size can be reduced. Considering such principles, the adaptation mechanism uses the age of the individuals and their solution quality to evaluate the category of the individuals produced by the optimization process.

In order to compute the age of each individual, it is assigned a counter $c_i(i \in (1,2,3,\ldots,N_p))$ to each element $\mathbf{m}_i$ of M(k). When the initial population M(k) is created, all the counters are set to zero. Since the memory M(k) is updated at each generation, some elements prevail and others will be substituted by new individuals. Therefore, the counter of the surviving elements is incremented by one whereas the counter of new added elements is set to zero.

Another important requirement to calculate the number of elements to be evolved is the solution quality provided by each individual. The idea is to identify two classes of elements, those that provide acceptable solutions and those that can be considered as unacceptable solutions. In order to classify each element, it is used the average fitness value J_A produced by all the elements of M(k). Thus, J_A is calculated as follows:

$$J_A = \frac{1}{N_p} \sum_{i=1}^{N_p} J(\mathbf{m}_i) \tag{7.4}$$

where $J(\cdot)$ represents the fitness value corresponding to $\mathbf{m}_i$. These values can be evaluated by using the true objective function $f(\mathbf{m}_i)$ or by using the fitness estimation strategy $F(\mathbf{m}_i)$. Considering the average fitness value, two groups can be build: the set G constituted by the elements of M(k) whose fitness values are greater than J_A and the set B which groups the elements of M(k) whose fitness values are equal or lower than J_A.

Therefore, the number of individuals of the current population that will be incremented or decremented at each generation is calculated by the following model:

$$A = \text{floor}\left(\frac{|\mathbf{G}| \cdot \sum_{l \in \mathbf{G}} c_l - |\mathbf{B}| \cdot \sum_{q \in \mathbf{B}} c_q}{s}\right) \tag{7.5}$$

where the floor$(\cdot)$ function maps a real number to the previous integer. $|\mathbf{G}|$ and $|\mathbf{B}|$ represent the number of elements of G and B, respectively whereas $\sum_{l\in\mathbf{G}} c_l$ and $\sum_{q\in\mathbf{B}} c_q$ indicate the sum of the counters that correspond to the elements of G and B, respectively. s is a constant factor that allows the fine tuning. Finally, after intensive experimentation, s has been set to 10.

Therefore, the number the elements that define the population to be evolved is computed according to the following model:

$$N_e^k = N_e^{k-1} - A \tag{7.6}$$

Since the value of A can be positive or negative, the size of the population $\mathbf{P}^k$ may be higher or lesser than $\mathbf{P}^{k-1}$. In order to maintain stable the algorithm's operation, the population size has been protected to remain within fixed limits $(4 \le N_e^k \le 2 \cdot N_p)$. The computational procedure that implements this method is presented in Algorithm 7.1, in form of pseudo code.

Algorithm 7.1. Selection of the number of individuals N_e^k to be evolved	
1:	Input: Current population M(k), counters $c_1, \ldots, c_{N_p}$, the past number of individuals N_e^{k-1} and the constant factor s.
2:	$J_A \leftarrow \frac{1}{N_p}\sum_{i=1}^{N_p} J(\mathbf{m}_i)$
3:	G $\leftarrow$ FindIndividualsOverJA(M(k), J_A)
4:	B $\leftarrow$ FindIndividualsUnderJA(M(k), J_A)
5:	$c_l \leftarrow$ FindCountersOfG(G) (Where $l \in \mathbf{G}$)
6:	$c_q \leftarrow$ FindCountersOfB(B) (Where $q \in \mathbf{B}$)
7:	$\text{A} \leftarrow \text{floor}\left(\frac{|\mathbf{G}|\cdot\sum_{l\in\mathbf{G}} c_l - |\mathbf{B}|\cdot\sum_{q\in\mathbf{B}} c_q}{s}\right)$
8:	$N_e^k \leftarrow N_e^{k-1} - A$
9:	$N_e^k = \begin{cases} 4 & \text{if } N_e^k < 4 \\ N_e^k & \text{if } 4 < N_e^k < 2\cdot N_p \\ 2\cdot N_p & \text{if } N_e^k > 2\cdot N_p \end{cases}$
10:	Output: The number N_e^k

7.2.2.2 Selection Strategy for Building $\mathbf{P}^k$

Once the number of individuals has been defined, the next step is to select the elements from M(k) for building . The first action is to generate a new population MO which contains the same elements that M(k), but sorted according to their fitness values. Thus, MO presents in its first positions the elements whose fitness values are better than those located in the last positions. Then, MO is divided in two parts: X and Y. The section X corresponds to the first elements of MO whereas the rest of the elements constitute the part Y. Figure 7.1 shows this process.

In order to promote diversity, in the selection strategy, the 80 % of the individuals of are taken from the first elements of X and named as Fe as shown in Fig. 7.2, where Fe = floor(0.8*). The remaining 20 % of the individuals are randomly selected from section Y. Hence, the last set of Se elements (where Se = −Fe) is chosen considering that all elements of Y have the same possibility of being selected. Figure 7.2 shows a

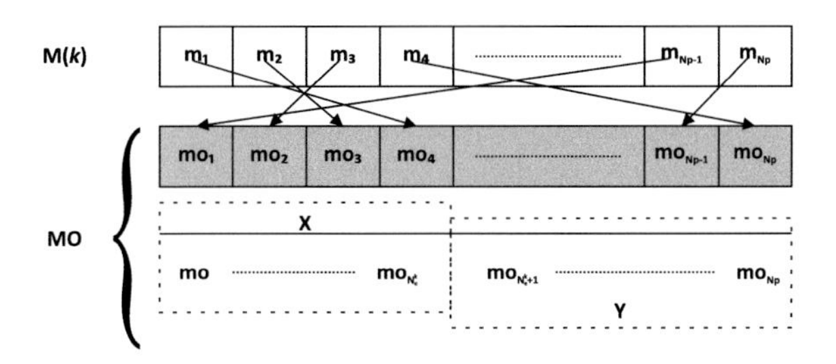

Fig. 7.1 Necessary post-processing implemented by the selection strategy

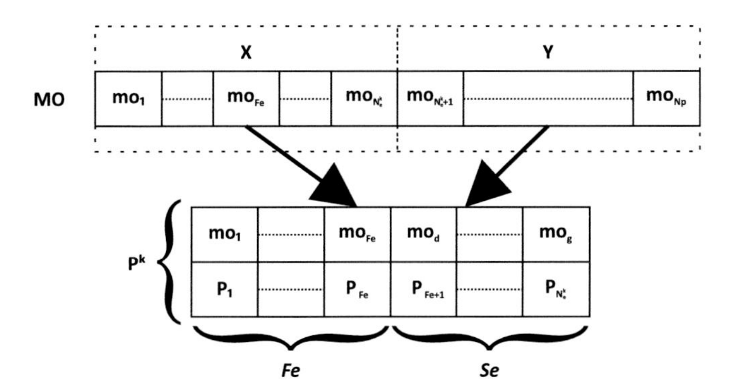

Fig. 7.2 Employed selection strategy to build the population $\mathbf{P}^k$, where $d, g \in \mathbf{Y}$

description of the selection strategy. The computational procedure that implements this method is presented in Algorithm 7.2, in form of pseudo code.

Algorithm 7.2 Selection strategy for building $\mathbf{P}^k$	
1:	Input: Current population M(k) and the number of individuals N_e^k.
2:	MO ← SortElementsFitness(M(k))
3:	[X,Y] ← DivideMO(MO, N_e^k)
4:	Fe ← floor(0.8* N_e^k)
5:	Se ← N_e^k - Fe
6:	$(\mathbf{p}_1^k, \quad ,\mathbf{p}_{Fe}^k)$ ← SelectElementsOfX(X, Fe)
7:	$(\mathbf{p}_{Fe+1}^k, \quad ,\mathbf{p}_{Fe+Se}^k)$ ← SelectRandomElementsOfY(Y, Se)
8:	Output: Population $\mathbf{P}^k$ to be evolve

7.2.3 *Seeking Operation*

The first main operation applied to the population $\mathbf{P}^k$ is the seeking operation. Considering $\mathbf{P}^k$ as the input population, APRE mutates $\mathbf{P}^k$ to produce a temporal population $\mathbf{T}^k$ of N_e^k vectors. In the seeking operation two different mutation models are used: the mutation employed by the Differential Evolution algorithm (DE) [21] and the trigonometric mutation operator [22].

7.2.3.1 DE Mutation Operator

In this mutation, three distinct individuals $r1$, $r2$, and $r3$ are randomly selected from the current population $\mathbf{P}^k$. Then, it is created a new value $h_{i,j}$ considering the following model:

$$h_{i,j} = p_{r1,j} + F(p_{r2,j} - p_{r3,j}) \tag{7.7}$$

where $r1$, $r2$, and $r3$ are randomly selected individuals such that they satisfy: $r1 \neq r2 \neq r3 \neq i$; $i = 1$ to N_e^k (population size), and $j = 1$ to n (number of decision variable). Hence, $p_{i,j}$ is the jth parameter of the ith individual of $\mathbf{P}^k$. The scale factor, F (0,1+), is a positive real number that controls the rate at which the population evolves.

7.2.3.2 Trigonometric Mutation Operator

The trigonometric mutation operation [22] is performed according to the following formulation:

$$\begin{aligned}
h_{i,j} &= p_{Av}(j) + F_1(p_{r1,j} - p_{r2,j}) + F_2(p_{r2,j} - p_{r3,j}) + F_3(p_{r3,j} - p_{r1,j}) \\
p_{Av}(j) &= \frac{p_{r1,j} + p_{r2,j} + p_{r3,j}}{3} \\
F_1 &= (d_{r2} - d_{r1}), F_2 = (d_{r3} - d_{r2}), F_3 = (d_{r1} - d_{r3}) \\
d_{r1} &= \frac{|J(\mathbf{p}_{r1})|}{d_T}, d_{r2} = \frac{|J(\mathbf{p}_{r2})|}{d_T}, d_{r3} = \frac{|J(\mathbf{p}_{r3})|}{d_T} \\
d_T &= |J(\mathbf{p}_{r1})| + |J(\mathbf{p}_{r2})| + |J(\mathbf{p}_{r3})|
\end{aligned} \tag{7.8}$$

where $\mathbf{p}_{r1}, \mathbf{p}_{r2}$ and $\mathbf{p}_{r3}$ represent the individuals $r1$, $r2$, and $r3$ randomly selected from the current population $\mathbf{P}^k$ whereas $J(\cdot)$ represents the fitness value (calculated or estimated) corresponding to $\mathbf{p}_i$. Under this formulation, the individual $p_{Av}(j)$ to be perturbed is the average value of three randomly selected vectors ($r1$, $r2$, and $r3$). The perturbation to be imposed over such individual is implemented by the sum of three weighted vector differentials. F_1, F_2, and F_3 are the weights applied to these vector differentials. It is noted that the trigonometric mutation is a greedy operator since it biases the $p_{Av}(j)$ strongly in the direction where the best one of three individuals is lying.

Computational Procedure

Considering $\mathbf{P}^k$ as the input population, all its N_e^k individuals are sequentially processed in cycles beginning by the first individual $\mathbf{p}_1$. Therefore, in the cycle i (where it is processed the individual i), three distinct individuals $r1$, $r2$, and $r3$ are randomly selected from the current population considering that they satisfy the following conditions $r1 \neq r2 \neq r3 \neq i$. Then, it is processed each dimension of $\mathbf{p}_i$ beginning by the first parameter 1 until the last dimension n has been reached. At each processing cycle, the parameter $p_{i,j}$ considered as a parent, creates an offspring $t_{i,j}$ in two steps. In the first step, from the selected individuals $r1$, $r2$, and $r3$, a donor vector $h_{i,j}$ is created by means of two different mutation models. In order to select which mutation model is applied, a uniform random number is generated within the range [0,1]. If such number is less than a threshold MR, the donor vector $h_{i,j}$ is generated by the DE mutation operator; otherwise, it is produced by the trigonometric mutation operator. Such process can be modeled as follows:

$$h_{i,j} = \begin{cases} \text{By using Eq.(7.7)} & \text{with probability } MR \\ \text{By using Eq.(7.8)} & \text{with probability } (1 - MR) \end{cases} \tag{7.9}$$

In the second step, the final value of the offspring $t_{i,j}$ is determined. Such decision is stochastic; hence a second uniform random number is generated within the range [0, 1]. If this random number is less than CH, $t_{i,j} = h_{i,j}$; otherwise, $t_{i,j} = p_{i,j}$. This operation can be formulated as following:

$$t_{i,j} = \begin{cases} h_{i,j} & \text{with probability } CH \\ p_{i,j} & \text{with probability } (1 - CH) \end{cases} \tag{7.10}$$

The complete computational procedure is presented in Algorithm 7.3, in form of pseudo code.

Algorithm 7.3. Seeking operation of APRE algorithm	
1:	Input: Current population $\mathbf{P}^k$
2:	for i=1 to N_e^k do
3:	$(\mathbf{p}_{r1},\mathbf{p}_{r2},\mathbf{p}_{r3}) \leftarrow$ SelectElements() %Considering that $r1 \neq r2 \neq r3 \neq i$
4:	for j=1 to n do
5:	if (rand(0,1)<=MR) then
6:	$h_{i,j} \leftarrow$ DEMutation $(\mathbf{p}_{r1},\mathbf{p}_{r2},\mathbf{p}_{r3})$ % Eq. 7.7
7:	else
8:	$h_{i,j} \leftarrow$ TrigonometricMutation $(\mathbf{p}_{r1},\mathbf{p}_{r2},\mathbf{p}_{r3})$ % Eq. 7.8
9:	end if
10:	if (rand(0,1)<=CH) then
11:	$t_{i,j} \leftarrow h_{i,j}$
12:	else
13:	$t_{i,j} \leftarrow p_{i,j}$
14:	end if
15:	end for
16:	end for
17:	Output: Population $\mathbf{T}^k$

7.2.4 Fitness Estimation Strategy

Once the population $\mathbf{T}^k$ has been generated by the seeking operation, it is necessary to calculate the fitness value provided by each individual. In order to reduce the number of function evaluations, it is introduced a fitness estimation strategy that decides which individuals can be estimated or actually evaluated. The idea of such strategy is to find the global optimum of a given function considering only a very few number of function evaluations.

In this chapter, we explore the use of a local approximation scheme that estimates the fitness values based on previously evaluated neighboring individuals, stored in the memory M(k) during the evolution process. The strategy decides if an individual $\mathbf{t}_i$ is calculated or estimated based on two criteria. The first one considers the distance between $\mathbf{t}_i$ and the nearest element $\mathbf{m}^{ne}$ contained in M(k) (where $\mathbf{m}^n \in (\mathbf{m}_1, \mathbf{m}_2, \ldots, \mathbf{m}_{N_p})$) whereas the second one examines the quality factor provided by the nearest element $\mathbf{m}^{ne}(Q(\mathbf{m}^{ne}))$.

In the model, individuals of $\mathbf{T}^k$ that are near the elements of M(k) holding the best quality values have a high probability to be evaluated. Such individuals are important, since they will have a stronger influence on the evolution process than other individuals. In contrast, individuals of $\mathbf{T}^k$ that are also near the elements of M(k) but with a bad quality value maintain a very low probability to be evaluated. Thus, most of such individuals will be only estimated, assigning it the same fitness value that the nearest element of M(k). On the other hand, it is also evaluated those individuals in regions of the search space with few previous evaluations (individuals of $\mathbf{T}^k$ located longer than a distance D). The fitness values of these individuals are uncertain; since there is no close reference (close points contained in M(k)) in order to calculate their estimates.

Therefore, the fitness estimation strategy follows two rules in order to evaluate or estimate the fitness values:

1. If the new individual $\mathbf{t}_i$ is located closer than a distance D with respect to the nearest element $\mathbf{m}^{ne}$ stored in M, then a uniform random number is generated within the range [0,1]. If such number is less than $Q(\mathbf{m}^{ne})$, $\mathbf{t}_i$ is evaluated by the true objective function $(f(\mathbf{t}_i))$; otherwise, it is estimated assigning it the same fitness value that $\mathbf{m}^{ne}$ $(F(\mathbf{t}_i) = J(\mathbf{m}^{ne}))$. Figure 7.3a, b draw the rule procedure.
2. If the new individual $\mathbf{t}_i$ is located longer than a distance D with respect to the nearest individual location $\mathbf{m}^{ne}$ stored in M, then the fitness value of $\mathbf{t}_i$ is evaluated using the true objective function $(f(\mathbf{t}_i))$. Figure 7.3c outlines the rule procedure.

From the rules, it is clear that the distance D controls the trade-off between the evaluation and estimation of new individuals. Unsuitable values of D result in a lower convergence rate, longer computation time, larger function evaluation number, convergence to a local maximum or unreliability of solutions. Therefore, the D value is computed considering the following procedure: First, the interval differences from all n decision variables are calculated $(d_j = p_j^{high} - p_j^{low}$, $j \in 1, 2, \ldots, n)$. Then, it is identified the minimal difference $d_{\min}$ among them. Finally, the distance D is computed considering following the model:

$$D = \frac{d_{\min}}{50} \tag{7.11}$$

The two rules show that the fitness estimation strategy is simple and straightforward. Figure 7.3 illustrates the procedure of fitness computation for a new candidate solution $\mathbf{t}_i$ considering the two different rules. In the problem the objective function f is maximized with respect to two parameters (x_1, x_2). In all figures (Fig. 7.3a–c) the memory M(k) contains five different elements $(\mathbf{m}_1, \mathbf{m}_2, \mathbf{m}_3, \mathbf{m}_4, \mathbf{m}_5)$ with their corresponding fitness values $(J(\mathbf{m}_1),\ J(\mathbf{m}_2),\ J(\mathbf{m}_3),\ J(\mathbf{m}_4),\ J(\mathbf{m}_5))$ and quality factors $(Q(\mathbf{m}_1),\ Q(\mathbf{m}_2),\ Q(\mathbf{m}_3),\ Q(\mathbf{m}_4),\ Q(\mathbf{m}_5))$. Figure 7.3a, b show the fitness evaluation $(f(x_1, x_2))$ or estimation $(F(x_1, x_2))$ of the new individual $\mathbf{t}_i$ following the rule 1.

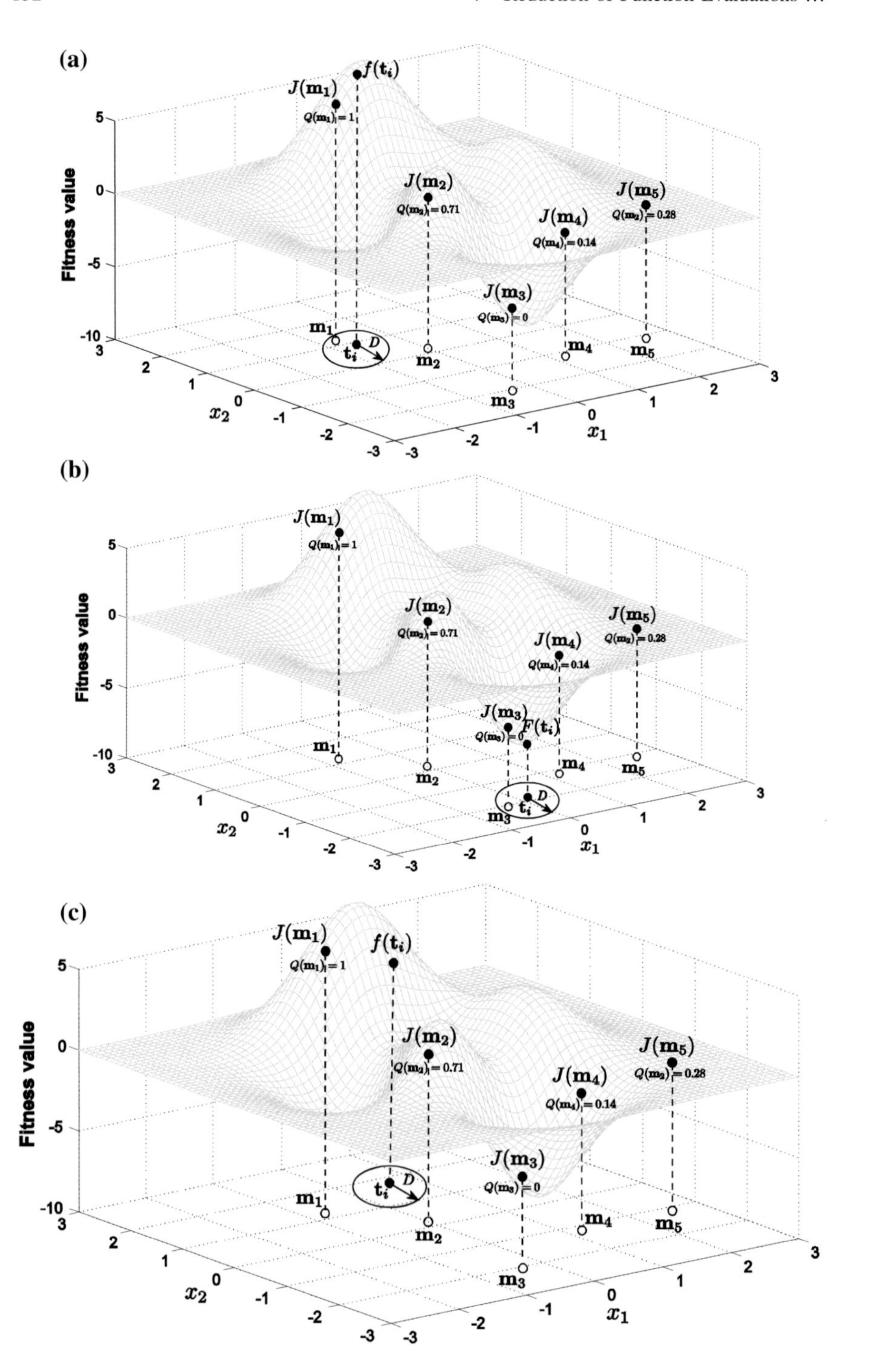
(a)
(b)
(c)
Fitness value
x_1
x_2
$J(\mathbf{m}_1)$
$J(\mathbf{m}_2)$
$J(\mathbf{m}_3)$
$J(\mathbf{m}_4)$
$J(\mathbf{m}_5)$
$Q(\mathbf{m}_1) = 1$
$Q(\mathbf{m}_2) = 0.71$
$Q(\mathbf{m}_3) = 0$
$Q(\mathbf{m}_4) = 0.14$
$Q(\mathbf{m}_2) = 0.28$
$f(\mathbf{t}_i)$
$F(\mathbf{t}_i)$
$\mathbf{t}_i$
D
$\mathbf{m}_1$
$\mathbf{m}_2$
$\mathbf{m}_3$
$\mathbf{m}_4$
$\mathbf{m}_5$

◀ **Fig. 7.3** The fitness estimation strategy. **a** According to the rule 1, the individual $\mathbf{t}_i$ has a high probability to be evaluated $f(\mathbf{t}_1)$, since it is located closer than a distance D with respect to the nearest element $\mathbf{m}^{ne} = \mathbf{m}_1$ whose quality factor $Q(\mathbf{m}_1)$ corresponds to the best value. **b** According to the rule 1, the individual $\mathbf{t}_i$ has a high probability to be estimated $F(\mathbf{t}_1)$ (assigning it the same fitness value that $\mathbf{m}^{ne}$ $(F(\mathbf{t}_i) = J(\mathbf{m}_3))$), since it is located closer than a distance D with respect to the nearest element $\mathbf{m}^{ne} = \mathbf{m}_3$ whose quality factor $Q(\mathbf{m}_3)$ corresponds to the worst value. **c** According to the rule 2, the individual $\mathbf{t}_i$ is evaluated, as there is no close reference in its neighborhood

Figure 7.3a represent the case when $\mathbf{m}^{ne}$ holds a good quality factor whereas Fig. 7.3b when $\mathbf{m}^{ne}$ maintains a bad quality factor. Finally, Fig. 7.3c presents the fitness evaluation of $\mathbf{t}_i$ considering the conditions of rule 2. The procedure that implements the fitness estimation strategy is presented in Algorithm 7.4, in form of pseudo code.

Algorithm 7.4. Fitness estimation strategy		
1:	Input: Population $\mathbf{T}^k$ and memory M(k)	
2:	for i=1 to N_e^k do	
3:	$\mathbf{m}^{ne} \leftarrow$ FindNearestElementOfM($\mathbf{t}_i$)	
4:	*distance* $\leftarrow$ FindTheDistance($\mathbf{t}_i$, $\mathbf{m}^{ne}$)	
5:	if (distance<D) then	
6:	if (rand(0,1)<=$Q(\mathbf{m}^{ne})$) then	Rule 1
7:	$J(\mathbf{t}_i) \leftarrow f(\mathbf{t}_i)$ % Evaluation	
8:	else	
9:	$J(\mathbf{t}_i) \leftarrow J(\mathbf{m}^{ne})$ % Estimation	
10:	end if	
11:	else	
12:	$J(\mathbf{t}_i) \leftarrow f(\mathbf{t}_i)$ % Evaluation	Rule 2
13:	end if	
14:	end for	
15:	Output: fitness values of $\mathbf{T}^k$	

7.2.5 Memory Updating

Once the operation of seeking and fitness estimation have been applied, it is necessary to update the memory M(k). In the APRE algorithm, the memory M(k) is updated considering the following procedure:

1. The elements of M(k) and $\mathbf{T}^k$ are merged into $\mathbf{M}_U(\mathbf{M}_U = \mathbf{M}(k) \cup \mathbf{T}^k)$.
2. From the resulting elements of $\mathbf{M}_U$, it is selected the N_p best elements according to their fitness values to build the new memory M(k + 1).

3. The counters $c_1, c_2, \ldots, c_{N_p}$ must be updated. Thus, the counter of the surviving elements is incremented by 1 whereas the counter of modified elements is set to zero.

7.2.6 Refinement Operation

The second main operation applied by the APRE algorithm is the refinement operation. Under this operation, it is improved the solution quality of existent promising solutions within a small neighborhood. In order to implement such process, it is generated a new memory ME which contains the same elements that M(k + 1), but sorted according to their fitness values. Thus, ME presents in its first positions the elements whose fitness values are better than those located in the last positions. Then, the 10 % of the $N_p(N_e)$ individuals are taken from the first elements of ME to build the set E ($\mathbf{E} = \{\mathbf{me}_1, \mathbf{me}_2, \ldots, \mathbf{me}_{N_e}\}$ where $N_e = \text{ceil}(0.1 \cdot N_p)$).

To each element $\mathbf{me}_i$ of E is assigned a probability p_i which express the likelihood of the element $\mathbf{me}_i$ to be refined. Such probability is computed as follows:

$$p_i = \frac{N_e + 1 - i}{N_e} \tag{7.12}$$

Therefore, the first elements of E have a better probability to be exploited than the last ones. In order to decide if the element $\mathbf{me}_i$ must be refined, a uniform random number is generated within the range [0,1]. If such number is less p_i, the element $\mathbf{me}_i$ will be modified by the refinement operation; otherwise, it remains without changes.

If the refinement operation over $\mathbf{me}_i$ is verified, the position of $\mathbf{me}_i$ is perturbed considering a small neighborhood. The idea is to test if it is possible to refine the solution provided by $\mathbf{me}_i$ modifying slightly its position. In order to improve the refinement process, APRE starts perturbing the original position within the interval [−D, D] (where D is the distance defined in Eq. 7.11) and then gradually is reduced as the process evolves. Thus, the perturbation over a generic element $\mathbf{me}_i$ is modeled as follows:

$$me_{i,j}^{new} = me_{i,j} + \left[D\frac{ng - k}{ng}\right] \cdot \text{rand}(-1, 1) \tag{7.13}$$

where k is the current iteration and ng is the total number of iterations from which consist the evolution process. Once $\mathbf{me}_i^{new}$ has been calculated, its fitness value is computed by using the true objective function $(J(\mathbf{me}_i^{new}) = f(\mathbf{me}_i^{new}))$. If $\mathbf{me}_i^{new}$ is better than $\mathbf{me}_i$ according to their fitness values, the value of $\mathbf{me}_i$ in the original memory $\mathbf{M}(k+1)$ is updated with $\mathbf{me}_i^{new}$, otherwise the memory $\mathbf{M}(k+1)$ remains without changes. The procedure that implements the refinement operation is presented in Algorithm 7.5, in form of pseudo code.

Algorithm 7.5. Refinement operation of APRE	
1:	Input: New memory $\mathbf{M}(k+1)$, current iteration k
2:	ME $\leftarrow$ SortElementsFitness($\mathbf{M}(k+1)$)
3:	$N_e \leftarrow$ ceil$(0.1 \cdot N_p)$
4:	E $\leftarrow$ SelectTheFirstElements(ME, N_e)
5:	for i=1 to N_e do
6:	$p_i \leftarrow (N_e+1-i)/\ N_e$
7:	if (rand(0,1)<= p_i) then
8:	for j=1 to n do
9:	$me_{i,j}^{new} \leftarrow me_{i,j} + \left[D \dfrac{ng-k}{ng} \right] \cdot \text{rand}(-1,1)$
10:	end for
11:	$J(\mathbf{me}_i^{new}) \leftarrow f(\mathbf{me}_i^{new})$
12:	if$\left(J(\mathbf{me}_i^{new}) > J(\mathbf{me}_i)\right)$ then
13:	$\mathbf{M}(k+1) \leftarrow$ MemoryIsUpdated($\mathbf{me}_i^{new}$)
14:	end if
15:	end if
16:	end for
17:	Output: Memory $\mathbf{M}(k+1)$

In order to demonstrate the refinement operation, Fig. 7.4a illustrates a simple example. It is assumed a memory $\mathbf{M}(k+1)$ of ten different 2-dimensional elements $(N_p = 10)$. Figure 7.4b shows the previous configuration of the presented example before the refinement operation takes place. Since only the 10 % of the best elements of $\mathbf{M}(k+1)$ will build the set E, $\mathbf{m}_5$ is the single element that constitutes E $(\mathbf{me}_1 = \mathbf{m}_5)$. Therefore, according to Eq. 7.12, the probability p_1 assigned to $\mathbf{me}_1$ is 1.

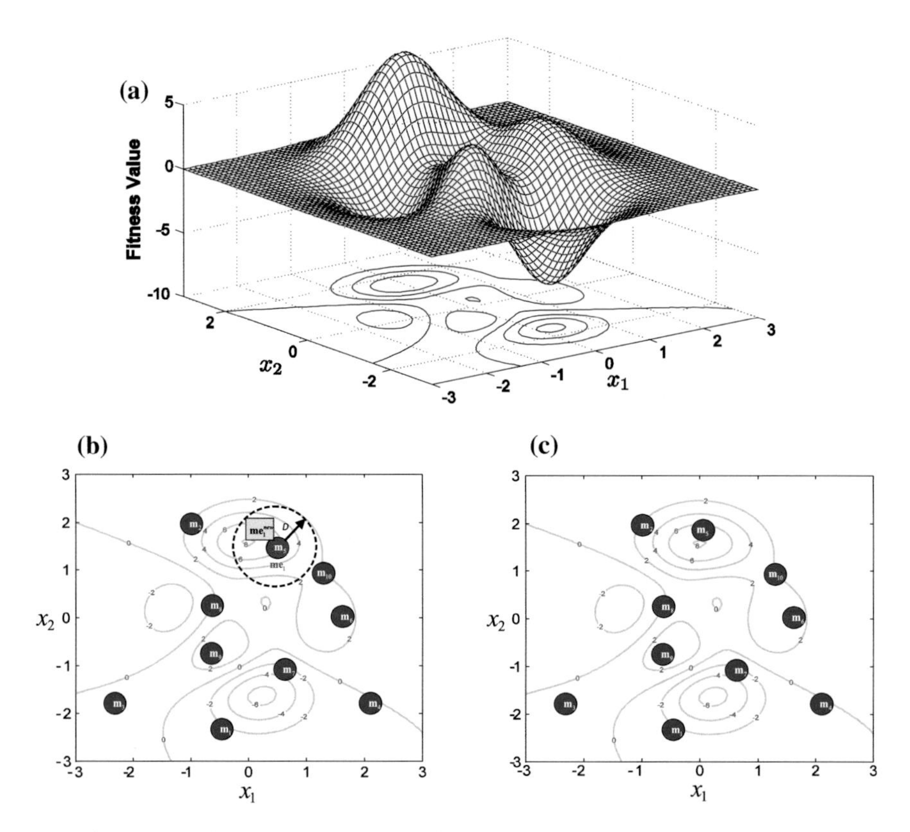

Fig. 7.4 Example of the mating operation: **a** function example, **b** initial configuration before the refinement operation and **c** configuration after the operation

Under such circumstances, the element $\mathbf{me}_1$ is perturbed considering the Eq. 7.13, generating the new position $\mathbf{me}_1^{new}$. As $\mathbf{me}_1^{new}$ is better than $\mathbf{me}_1$ according to their fitness values, the value of $\mathbf{m}_5$ in the original memory $\mathbf{M}(k+1)$ is updated with $\mathbf{me}_i^{new}$. Figure 7.4c shows the final configuration of $\mathbf{M}(k+1)$ after the refinement operation has been achieved.

7.2.6.1 Computational Procedure

The computational procedure for APRE can be summarized as follows:

The APRE algorithm is an iterative process in which several actions are executed. After initialization (lines 2–3), it is computed the number of memory

Computational procedure of APRE	
1:	Input: N_p, N_e^0, *MR*, *CH* and Max*k* (where Max*k* is the maximum number of iterations).
2:	M(1) $\leftarrow$ InitializeM(N_p)
3:	$c_1,\quad ,c_{N_p}$ $\leftarrow$ ClearCounters( )
4:	for *k*=1 to Max*k* do
5:	Algorithm 1 (Section 2.2.1)
6:	Algorithm 2 (Section 2.2.2)
7:	Algorithm 3 (Section 2.3)
8:	Algorithm 4 (Section 2.4)
9:	M(*k*+1) $\leftarrow$ UpdateM(M(*k*)) (Section 2.5)
10:	$c_1,\quad ,c_{N_p}$ $\leftarrow$ UpdateCounters $\left(c_1,\quad ,c_{N_p}\right)$
11:	Q $\leftarrow$ CalculateQualityFactor(M(*k*))
12:	Algorithm 5
13:	end for
14:	*Solution* $\leftarrow$ FindBestElement(M(*k*))
15:	Output: *Solution*

elements to be evolved. Such number is automatically modified in each iteration (lines 5–6). Then, a set of new individuals is generated as a consequence of the execution of the seeking operation (line 7). For each new individual, its fitness value is estimated or evaluated according to a decision taken by a fitness estimation strategy (line 8). Afterwards, the memory is updated. For this process, the new individuals produced by the seeking operation compete against the memory elements to build the final memory configuration (lines 9–11). Finally, a sample of the best elements contained in the final memory configuration is undergone to the refinement operation (line 12). This cycle is repeated until the maximum number the iterations Max k has been reached.

7.3 Experimental Results

7.3.1 Test Suite and Experimental Setup

A comprehensive set of 23 functions, collected from Refs. [23–33], have been used to test the performance of APRE. Tables 7.1, 7.2, 7.3 and 7.4 present the benchmark functions used in our experimental study. Such functions are classified into four different categories: Unimodal test functions (Table 7.1), multimodal test functions (Table 7.2) and multimodal test functions with fixed dimensions (Tables 7.3 and 7.4). In such tables, n is the dimension of function, f_{opt} is the minimum value of the function and S is a subset of R^n. The optimum location $(\mathbf{x}_{opt})$ for functions in Tables 7.1 and 7.2 fall into $[0]^n$, except for f_5,f_{12} and f_{13} with $\mathbf{x}_{opt}$ falling into $[1]^n$ and f_8 in $[420.96]^n$.

Table 7.1 Unimodal test functions

Test function	S	f_{opt}
$f_1(\mathbf{x}) = \sum_{i=1}^{n} x_i^2$	$[-100, 100]^n$	0
$f_2(\mathbf{x}) = \sum_{i=1}^{n} \lvert x_i \rvert + \prod_{i=1}^{n} \lvert x_i \rvert$	$[-10, 10]^n$	0
$f_3(\mathbf{x}) = \sum_{i=1}^{n} \left(\sum_{j=1}^{i} x_j\right)^2$	$[-100, 100]^n$	0
$f_4(\mathbf{x}) = \max_i \{\lvert x_i \rvert, 1 \le i \le n\}$	$[-100, 100]^n$	0
$f_5(\mathbf{x}) = \sum_{i=1}^{n-1} \left[100(x_{i+1} - x_i^2)^2 + (x_i - 1)^2\right]$	$[-30, 30]^n$	0
$f_6(\mathbf{x}) = \sum_{i=1}^{n} (x_i + 0.5)^2$	$[-100, 100]^n$	0
$f_7(\mathbf{x}) = \sum_{i=1}^{n} ix_i^4 + rand(0, 1)$	$[-1.28, 1.28]^n$	0

7.3.2 Performance Comparison with Other Metaheuristic Approaches

The APRE algorithm has been applied to 23 test functions in order to compare its performance with other approaches that have been conceived to reduce the number of function evaluations. The methods considered in the experiments include the CSA-Nearest neighbors (CSA-NN) [8], the PSO-radial basis functions (PSO-RBF) [9] and the ES-kriging method (ES-K) [10].

CSA-NN conducts a search strategy considering the principles of the clonal selection algorithm. In order to reduce the number of function evaluations, CSA-NN uses an approximated model based on the nearest neighbors schema. On the other hand, PSO-RBF considers multiple trial positions for each particle in the swarm and uses a surrogate model to identify the most promising trial position. Therefore, the current overall best position is refined by finding the global minimum of the surrogate in the neighborhood of that position. In the method, the approximated model is implemented by using a set of radial basis functions (RBF). Finally, the ES-K algorithm employs a common evolutionary strategy to produce new candidate solutions. The approach incorporates a classification system to predict which candidate solution should be evaluated or estimated in the evolution process. In case of estimation, the candidate solution is assessed by using a kriging model that approximates the original fitness function.

In the comparison, the population size is set to 50 for all cases, except for APRE in which such parameter is adapted as the process evolves. The maximum iteration number is 1000 for functions in Tables 7.1 and 7.2 and 500 for functions in Table 7.3. Such stop criteria have been chosen as to keep compatibility to similar works which are reported in [23–33]. CSA-NN, PSO-RBF and ES-K have been configured, as it is recommended by their respective references. On the other hand, after extensive experimentation, the APRE algorithm has been set as follows: $N_p = 50, N_e^0 = 20$, $MR = 0.6$ and $CH = 0.8$. Once these parameters have been fixed, they are kept for all experiments in this section.

Table 7.2 Multimodal test functions

Test function	S	f_{opt}
$f_8(\mathbf{x}) = \sum_{i=1}^{n} -x_i \sin\left(\sqrt{\lvert x_i \rvert}\right)$	$[-500, 500]^n$	$[420.96]^n$
$f_9(\mathbf{x}) = \sum_{i=1}^{n} \left[x_i^2 - 10\cos(2\pi x_i) + 10\right]$	$[-5.12, 5.12]^n$	0
$f_{10}(\mathbf{x}) = -20\exp\left(-0.2\sqrt{\frac{1}{n}\sum_{i=1}^{n} x_i^2}\right) - \exp\left(\frac{1}{n}\sum_{i=1}^{n}\cos(2\pi x_i)\right) + 20$	$[-32, 32]^n$	0
$f_{11}(\mathbf{x}) = \frac{1}{4000}\sum_{i=1}^{n} x_i^2 - \prod_{i=1}^{n}\cos\left(\frac{x_i}{\sqrt{i}}\right) + 1$	$[-600, 600]^n$	0
$f_{12}(\mathbf{x}) = \frac{\pi}{n}\left\{10\sin(\pi y_1) + \sum_{i=1}^{n-1}(y_i - 1)^2\left[1 + 10\sin^2(\pi y_{i+1})\right] + (y_n - 1)^2\right\} + \cdots \sum_{i=1}^{n} u(x_i, 10, 100, 4)$ $y_i = 1 + \frac{x_i + 1}{4} \quad u(x_i, a, k, m) = \begin{cases} k(x_i - a)^m & x_i > a \\ 0 & -a < x_i < a \\ k(-x_i - a)^m & x_i < -a \end{cases}$	$[-50, 50]^n$	0
$f_{13}(\mathbf{x}) = 0.1\left\{\sin^2(3\pi x_1) + \sum_{i=1}^{n}(x_i - 1)^2\left[1 + \sin^2(3\pi x_i + 1)\right] \cdots + (x_n - 1)^2\left[1 + \sin^2(2\pi x_n)\right]\right\} + \sum_{i=1}^{n} u(x_i, 5, 100, 4)$	$[-50, 50]^n$	0

Table 7.3 Multimodal test functions with fixed dimensions

Test function	S	f_{opt}
$f_{14}(\mathbf{x}) = \left(\frac{1}{500} + \sum_{j=1}^{25} \frac{1}{j + \sum_{i=1}^{2} (x_i - a_{ij})^6} \right)^{-1}$ $a_{ij} = \begin{pmatrix} -32, -16, 0, 16, 32, -32, \ldots, 0, 16, 32 \\ -32, -32, -32, -32, 16, \ldots, 32, 32, 32 \end{pmatrix}$	$[-65.5, 65.5]^2$	1
$f_{15}(\mathbf{x}) = \sum_{i=1}^{11} \left[a_i - \frac{x_i(b_i^2 + b_i x_2)}{b_i^2 + b_i x_3 + x_4} \right]^2$ $\mathbf{a} = [0.1957, 0.1947, 0.1735, 0.1600, 0.0844, 0.0627,$ $0.0456, 0.0342, 0.0342, 0.0235, 0.0246]$ $\mathbf{b} = [0.25, 0.5, 1, 2, 4, 6, 8, 10, 12, 14, 16]$	$[-5, 5]^4$	0
$f_{16}(x_1, x_2) = 4x_1^2 - 2.1x_1^4 + \frac{1}{3}x_1^6 + x_1 x_2 - 4x_2^2 + 4x_2^4$	$[-5, 5]^2$	−1.0316
$f_{17}(x_1.x_2) = \left(x_2 - \frac{5.1}{4\pi^2}x_1^2 + \frac{5}{\pi}x_1 - 6\right)^2 + 10\left(1 - \frac{1}{8\pi}\right)\cos x_1 + 10$	$x_1 \in [-5, 10]$ $x_2 \in [0, 15]$	0.398
$f_{18}(x_1, x_2) = \left[1 + (x_1 + x_2 + 1)^2 (19 - 14x_1 + 3x_1^2 - 14x_2 + 6x_1 x_2 + 3x_2^2\right] \cdots$ $\times \left[30 + (2x_1 - 3x_2)^2 \times (18 - 32x_1 + 12x_1^2 + 48x_2 - 36x_1 x_2 + 27x_2^2\right]$	$[-5, 5]^2$	−3.87
$f_{19}(\mathbf{x}) = -\sum_{i=1}^{4} c_i \exp\left(-\sum_{j=1}^{3} a_{ij}(x_j - p_{ij})^2\right)$ $\mathbf{a} = \begin{bmatrix} 3 & 10 & 30 \\ 0.1 & 10 & 35 \\ 3 & 10 & 30 \\ 0.1 & 10 & 30 \end{bmatrix}$ $\mathbf{c} = [1, 1.2, 3, 3.2]$ $\mathbf{P} = \begin{bmatrix} 0.3689 & 0.117 & 0.2673 \\ 0.4699 & 0.4387 & 0.7470 \\ 0.1091 & 0.8732 & 0.5547 \\ 0.0381 & 0.5743 & 0.8828 \end{bmatrix}$	$[0, 1]^3$	−3.86

(continued)

Table 7.3 (continued)

Test function	S	f_{opt}
$f_{20}(\mathbf{x}) = -\sum_{i=1}^{4} c_i \exp\left(-\sum_{j=1}^{6} a_{ij}(x_j - p_{ij})^2\right)$ $\mathbf{a} = \begin{bmatrix} 10 & 3 & 17 & 3.5 & 1.7 & 8 \\ 0.05 & 10 & 17 & 0.1 & 8 & 14 \\ 3 & 3.5 & 17 & 10 & 17 & 8 \\ 17 & 8 & 0.05 & 10 & 0.1 & 14 \end{bmatrix}$ $\mathbf{c} = [1, 1.2, 3, 3.2]$ $\mathbf{P} = \begin{bmatrix} 0.131 & 0.169 & 0.556 & 0.012 & 0.828 & 0.588 \\ 0.232 & 0.413 & 0.830 & 0.373 & 0.100 & 0.999 \\ 0.234 & 0.141 & 0.352 & 0.288 & 0.304 & 0.665 \\ 0.404 & 0.882 & 0.873 & 0.574 & 0.109 & 0.038 \end{bmatrix}$	$[0, 1]^6$	−3.32
$f_{21}(\mathbf{x}) = -\sum_{i=1}^{5} \left[(\mathbf{x} - a_i)(\mathbf{x} - a_i)^T + c_i\right]^{-1}$ $\mathbf{a} = \begin{bmatrix} 4 & 4 & 4 & 4 \\ 1 & 1 & 1 & 1 \\ 8 & 8 & 8 & 8 \\ 6 & 6 & 6 & 6 \\ 3 & 7 & 3 & 7 \\ 2 & 9 & 2 & 9 \\ 5 & 5 & 3 & 3 \\ 8 & 1 & 8 & 1 \\ 6 & 2 & 6 & 2 \\ 7 & 3.6 & 7 & 3.6 \end{bmatrix}$ $\mathbf{c} = [0.1, 0.2, 0.2, 0.4, 0.4, 0.6, 0.3, 0.7, 0.5, 0.5]$	$[0, 10]^4$	−0.1532
$f_{22}(\mathbf{x}) = -\sum_{i=1}^{7} \left[(\mathbf{x} - a_i)(\mathbf{x} - a_i)^T + c_i\right]^{-1}$ a and c, equal to f_{21}	$[0, 10]^4$	−10.4028
$f_{23}(\mathbf{x}) = -\sum_{i=1}^{7} \left[(\mathbf{x} - a_i)(\mathbf{x} - a_i)^T + c_i\right]^{-1}$ a and c, equal to f_{21}	$[0, 10]^4$	−10.5363

Table 7.4 Optimum locations of Table 7.3

Test function	$\mathbf{x}_{opt}$	Test function	$\mathbf{x}_{opt}$
f_{14}	(−32, 32)	f_{19}	(0.114, 0.556, 0.852)
f_{15}	(0.1928, 0.1908, 0.1231, 0.1358)	f_{20}	(0.201, 0.15, 0.477, 0.275, 0.311, 0.657)
f_{16}	(0.089, −0.71), (−0.0089, 0.712)	f_{21}	5 local minima in a_{ij}, $i = 1,\ldots,5$
f_{17}	(−3.14, 12.27), (3.14, 2.275), (9.42, 2.42)	f_{22}	7 local minima in a_{ij}, $i = 1,\ldots,7$
f_{18}	(0, −1)	f_{23}	10 local minima in a_{ij}, $i = 1,\ldots,10$

Several experimental tests have been developed for comparing the performance of the APRE algorithm. The experiments have been developed considering the following function types:

- Uni-modal test functions (Table 7.1)
- Multimodal test functions (Table 7.2)
- Multimodal test functions with fixed dimensions (Tables 7.3 and 7.4)

7.3.2.1 Uni-modal Test Functions

In this test, the performance of the APRE algorithm is compared to CSA-NN, PSO-RBF and ES-K, considering functions with only one minimum/maximum. Such function type is represented by functions f_1 to f_7 in Table 7.1. The results, over 100 runs, are reported in Table 7.5 considering the following performance indexes: the average best-so-far solution, the average mean fitness function, the median of the best solution in the last iteration and the number of function evaluations. The best result for each function is boldfaced. According to Table 7.5, the APRE algorithm provides better results than CSA-NN, PSO-RBF and ES-K for all functions in terms of accuracy (the average best-so-far solution). On the other hand, the APRE method also employs a smaller number of function evaluations in comparison with the other algorithms on the majority of the test cases. Such an exception is presented in function f_5 where the ES-K method maintains a better performance. Most of the functions of Table 7.1 maintain a narrow curving valley that is hard to optimize, in case the search space cannot be explored properly [34]. For this reason, the performance differences are directly related to a better trade-off between exploration and exploitation that is produced by APRE operators (Fig. 7.5).

Moreover, the good convergence rate of APRE can be observed from Fig. 7.6. According to this figure, the APRE algorithm tends to find the global optimum faster than other algorithms and yet offering the highest convergence rate.

Table 7.5 Minimization result of benchmark functions in Table 7.1 with $n = 30$. Maximum number of iterations = 1000

		CSA-NN	PSO-RBF	ES-K	APRE
f_1	Average best so-far	3.1×10^{-3}	0.15	7.9×10^{-2}	$\mathbf{1.5\times10^{-5}}$
	Median best so-far	4.5×10^{-3}	0.21	9.5×10^{-2}	$\mathbf{1.1\times10^{-5}}$
	Average mean fitness	3.8×10^{-3}	0.17	8.9×10^{-2}	$\mathbf{1.4\times10^{-5}}$
	Number of function evaluations	42,141	30,541	24,878	21,954
f_2	Average best so-far	1.8×10^{-4}	0.02	4.21×10^{-3}	$\mathbf{3.11\times10^{-5}}$
	Median best so-far	2.7×10^{-4}	0.05	7.54×10^{-3}	$\mathbf{4.34\times10^{-5}}$
	Average mean fitness	2.0×10^{-4}	0.03	5.24×10^{-3}	$\mathbf{1.07\times10^{-5}}$
	Number of function evaluations	44,005	39,258	23,521	20,705
f_3	Average best so-far	1.25×10^{-3}	0.27	5.2×10^{-2}	$\mathbf{1.07\times10^{-4}}$
	Median best so-far	3.70×10^{-3}	0.32	7.2×10^{-2}	$\mathbf{1.78\times10^{-4}}$
	Average mean fitness	2.98×10^{-3}	0.28	6.2×10^{-2}	$\mathbf{0.87\times10^{-2}}$
	Number of function evaluations	32,875	28,479	25,652	22,007
f_4	Average best so-far	4.24×10^{-3}	0.27	3.42×10^{-2}	7.48×10^{-5}
	Median best so-far	7.11×10^{-3}	0.37	5.03×10^{-2}	6.07×10^{-5}
	Average mean fitness	5.16×10^{-3}	0.30	3.87×10^{-2}	1.10×10^{-4}
	Number of function evaluations	41,214	35,981	22,157	21,147
f_5	Average best so-far	1.29×10^{-2}	1.02	0.02	$\mathbf{4.21\times10^{-4}}$
	Median best so-far	3.71×10^{-2}	2.54	0.08	$\mathbf{1.02\times10^{-3}}$
	Average mean fitness	2.28×10^{-2}	1.78	0.05	$\mathbf{3.78\times10^{-3}}$
	Number of function evaluations	41,271	35,974	22,153	22,532
f_6	Average best so-far	1.18×10^{-4}	0.78×10^{-3}	3.81×10^{-4}	$\mathbf{1.14\times10^{-5}}$
	Median best so-far	1.78×10^{-4}	0.89×10^{-3}	5.41×10^{-4}	$\mathbf{1.97\times10^{-5}}$
	Average mean fitness	1.57×10^{-4}	0.81×10^{-3}	4.12×10^{-4}	$\mathbf{5.41\times10^{-5}}$
	Number of function evaluations	45,054	40,127	26,143	22,076
f_7	Average best so-far	1.25×10^{-3}	2.21×10^{-2}	5.34×10^{-3}	$\mathbf{7.01\times10^{-5}}$
	Median best so-far	3.47×10^{-3}	3.65×10^{-2}	8.11×10^{-3}	$\mathbf{2.10\times10^{-5}}$
	Average mean fitness	2.17×10^{-3}	3.07×10^{-2}	6.65×10^{-3}	$\mathbf{1.24\times10^{-4}}$
	Number of function evaluations	40,874	39,571	25,879	21,173

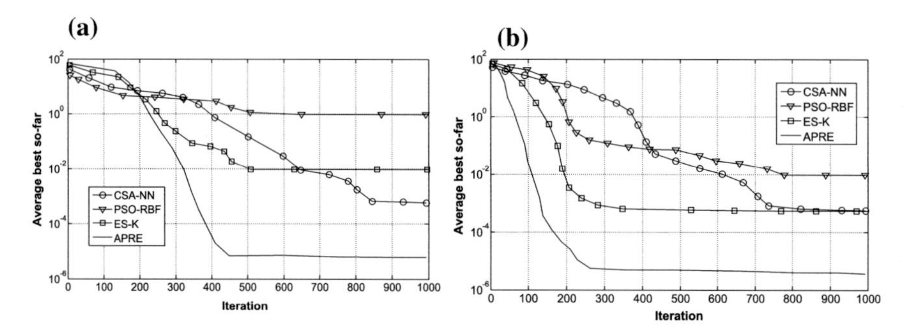

Fig. 7.5 Performance comparison of CSA-NN, PSO-RBF, ES-K and the APRE algorithm for minimization of **a** f_1 and **b** f_7 considering $n = 30$

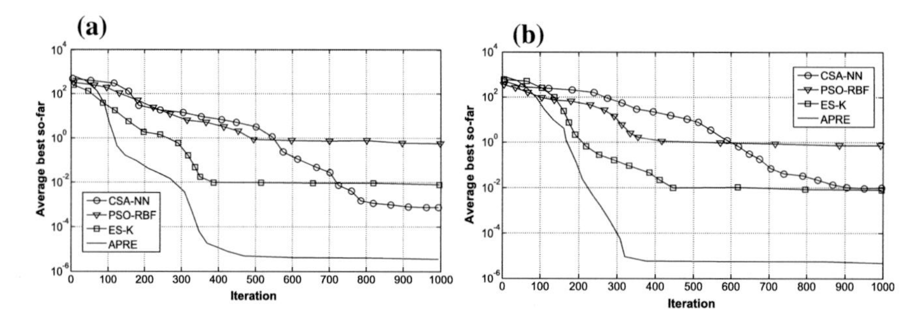

Fig. 7.6 Performance comparison of CSA-NN, PSO-RBF, ES-K and APRE for minimization of **a** f_{10} and **b** f_{12} considering $n = 30$

In order to statistically analyse the results in Table 7.4, it has been conducted a non-parametric significance proof known as Wilcoxon rank-sum test [35, 36] which is used to determine/verify if two algorithms are different statistically significant or not. In this method, the results of each algorithm are compared to calculate and analyze their differences. As a null hypothesis, it is assumed that there is no significant difference between mean values of the two algorithms. The alternative hypothesis considers a significant difference between both approaches. Technically speaking, this statistical test returns a parameter called p-value which determines the significance level of two algorithms. Under such circumstances, the difference is statistically significant if and only if it results in a p-value less than 0.05. Tables 7.6 and 7.7 report the p-values produced by Wilcoxon's test for the pair-wise comparison of the "average best so-far" and "number of function evaluations" of four groups. Such groups are constructed by APRE versus CSA-NN, APRE versus PSO-RBF and APRE versus ES-K. Most of the p-values reported in the tables are less than 0.05 (5 % significance level) which is a strong evidence against the null hypothesis, indicating that the APRE results are statistically significant and that it has not occurred by coincidence (i.e. due to the normal noise contained in the

Table 7.6 p-values produced by Wilcoxon's test comparing APRE versus CSA-NN, PSO-RBF and ES-K over the "average best-so-far" values from Table 7.5

APRE versus	CSA-NN	PSO-RBF	ES-K
f_1	0.031	0.004	0.014
f_2	0.067	0.001	0.018
f_3	0.061	0.007	0.011
f_4	0.021	0.002	0.015
f_5	0.034	0.003	0.016
f_6	0.041	0.005	0.011
f_7	0.024	0.001	0.017

Table 7.7 p-values produced by Wilcoxon's test comparing APRE versus CSA-NN, PSO-RBF and ES-K over the "number of function evaluations" values from Table 7.5

APRE versus	CSA-NN	PSO-RBF	ES-K
f_1	0.001	0.014	0.041
f_2	0.008	0.018	0.039
f_3	0.003	0.020	0.032
f_4	0.004	0.011	0.074
f_5	0.002	0.015	0.081
f_6	0.007	0.021	0.045
f_7	0.005	0.016	0.037

process). From Table 7.6, it is clear that the p-values of functions f_2 and f_3 in the group APRE versus CSA-NN are higher than 0.05. Such results reveal that there is not statistically difference in terms of precision between both methods for the involved functions. On the other hand, in Table 7.7, it is exhibited that there is not statistically difference with regard to the number of function evaluations between APRE and ES-K for functions f_4 and f_5. These results show that the presented method achieves the best balance over other algorithms, in terms of both accuracy and number of function evaluations.

7.3.2.2 Multimodal Test Functions

Multimodal functions, in contrast to unimodal, have many local minima/maxima which are, in general, more difficult to optimize. In this section the performance of the APRE algorithm is compared to other approaches considering multimodal functions. Such comparison reflects the algorithm's ability to escape from poor local optima and to locate a near-global optimum. The experiments have been done on f_8 to f_{13} of Table 7.2 where the number of local minima increases exponentially as the dimension of the function increases. The dimension of these functions is set to 30. The results are averaged over 100 runs, reporting the performance indexes in Table 7.8 as follows: the average best-so-far solution, the average mean fitness function and the median of the best solution in the last iteration and the number of function evaluations (the best result for each function is highlighted) Likewise, p-values of the Wilcoxon signed-rank test of 100 independent runs are listed in

Table 7.8 Minimization of benchmark functions in Table 7.2 with n = 30

		CSA-NN	PSO-RBF	ES-K	APRE
f_8	Average best so-far	$\mathbf{-1.26 \times 10^4}$	-1.04×10^4	-1.07×10^4	$\mathbf{-1.26 \times 10^4}$
	Median best so-far	-1.20×10^4	-1.02×10^4	-1.10×10^4	$\mathbf{-1.24 \times 10^4}$
	Average mean fitness	-1.07×10^4	-1.01×10^3	-1.04×10^3	$\mathbf{-1.21 \times 10^4}$
	Number of function evaluations	35,874	27,536	20,875	18,874
f_9	Average best so-far	1.1×10^{-3}	0.84	6.2×10^{-2}	$\mathbf{0.80 \times 10^{-3}}$
	Median best so-far	4.2×10^{-2}	1.21	8.7×10^{-2}	$\mathbf{1.10 \times 10^{-3}}$
	Average mean fitness	2.8×10^{-2}	0.97	7.0×10^{-2}	$\mathbf{1.18 \times 10^{-3}}$
	Number of function evaluations	40,874	30,257	27,387	21,087
f_{10}	Average best so-far	4.37×10^{-3}	0.14	3.12×10^{-2}	$\mathbf{2.11 \times 10^{-5}}$
	Median best so-far	8.21×10^{-3}	0.28	5.83×10^{-2}	$\mathbf{6.37 \times 10^{-5}}$
	Average mean fitness	6.14×10^{-3}	0.21	4.69×10^{-2}	$\mathbf{0.34 \times 10^{-4}}$
	Number of function evaluations	37,714	30,874	18,747	18,986
f_{11}	Average best so-far	1.08×10^{-2}	0.001	6.37×10^{-2}	$\mathbf{1.02 \times 10^{-5}}$
	Median best so-far	2.87×10^{-2}	0.004	9.04×10^{-2}	$\mathbf{2.14 \times 10^{-5}}$
	Average mean fitness	1.51×10^{-2}	0.002	7.53×10^{-2}	$\mathbf{0.97 \times 10^{-4}}$
	Number of function evaluations	34,287	30,127	23,874	21,003
f_{12}	Average best so-far	0.25×10^{-2}	0.08	6.74×10^{-2}	$\mathbf{2.2 \times 10^{-5}}$
	Median best so-far	0.61×10^{-2}	0.11	9.36×10^{-2}	$\mathbf{5.1 \times 10^{-5}}$
	Average mean fitness	0.48×10^{-2}	0.09	7.11×10^{-3}	$\mathbf{6.8 \times 10^{-5}}$
	Number of function evaluations	36,428	31,248	24,184	21,879
f_{13}	Average best so-far	6.87×10^{-6}	1.02×10^{-4}	3.5×10^{-5}	$\mathbf{0.98 \times 10^{-6}}$
	Median best so-far	8.54×10^{-6}	3.42×10^{-4}	5.8×10^{-5}	$\mathbf{1.1 \times 10^{-6}}$
	Average mean fitness	7.02×10^{-6}	2.78×10^{-4}	4.2×10^{-4}	$\mathbf{0.97 \times 10^{-6}}$
	Number of function evaluations	35,748	28,591	23,897	22,989

Maximum number of iterations = 1000

Tables 7.9 and 7.10, for the average best so-far and number of function evaluations data, respectively.

According to Table 7.8, for most of the functions, APRE yields better solutions and maintains a lower number of function evaluations than the others methods. The Wilcoxon rank test results, presented in Table 7.9, show that the APRE algorithm

Table 7.9 p-values produced by Wilcoxon's test comparing APRE versus CSA-NN, PSO-RBF and ES-K over the "average best-so-far" values from Table 7.8

APRE versus	CSA-NN	PSO-RBF	ES-K
f_8	0.087	0.024	0.002
f_9	0.001	0.012	0.041
f_{10}	0.014	0.017	0.011
f_{11}	0.011	0.016	0.009
f_{12}	0.010	0.021	0.007
f_{13}	0.078	0.010	0.041

Table 7.10 p-values produced by Wilcoxon's test comparing APRE versus CSA-NN, PSO-RBF and ES-K over the "number of function evaluations" values from Table 7.8

APRE versus	CSA-NN	PSO-RBF	ES-K
f_8	0.008	0.014	0.012
f_9	0.001	0.011	0.041
f_{10}	0.007	0.016	0.101
f_{11}	0.005	0.013	0.043
f_{12}	0.003	0.012	0.038
f_{13}	0.007	0.018	0.184

performed better than CSA-NN, PSO-RBF and ES-K considering most of the functions, whereas, from a statistical viewpoint, there is not statistically difference in terms of precision between the APRE algorithm and CSA-NN for functions f_8 and f_{13}. On the other hand, Table 7.6 shows that the APRE approach uses a significantly low number of function evaluations when it is statistically compared with its competitors. However, for functions f_{10} and f_{13}, it is exhibited that there is not statistically difference with regard to the number of function evaluations between APRE and ES-K. These results show that the presented method achieves the best balance over other algorithms, in terms of both accuracy and number of function evaluations. In order to visualize the convergence properties, Fig. 7.6 shows the evolution of the "average best-so-far" solutions for functions f_{10} and f_{12}.

7.3.2.3 Multimodal Test Functions with Fixed Dimensions

In the following experiments the performance of the APRE algorithm is compared to CSA-NN, PSO-RBF and ES-K considering functions which are extensively reported in the literature [22–33]. Such functions, represented by f_{14} to f_{23} in Tables 7.3 and 7.4, are all multimodal with fixed dimensions. Since such functions have a low dimensional profile, they do not represent a complex challenge [37]. Table 7.11 shows the outcome of the optimization process. Results show how all algorithms maintain a similar solution accuracy. However, there exist a strong differences among APRE and the other methods when they are compared in terms of the number of function evaluations. Figure 7.7 shows the optimization process for the functions f_{15} and f_{22}.

Table 7.11 Minimization result of benchmark functions in Table 7.3 with n = 30

		CSA-NN	PSO-RBF	ES-K	APRE
f_{14}	Average best so-far	0.998	0.998	0.998	0.998
	Median best so-far	0.990	0.988	0.985	0.995
	Average mean fitness	0.974	0.981	0.980	0.994
	Number of function evaluations	17,541	15,841	12,897	10,574
f_{15}	Average best so-far	3.4×10^{-4}	1.2×10^{-3}	4.8×10^{-4}	1.2×10^{-5}
	Median best so-far	5.2×10^{-4}	2.8×10^{-3}	9.2×10^{-4}	4.7×10^{-5}
	Average mean fitness	4.3×10^{-4}	1.7×10^{-3}	5.7×10^{-4}	0.4×10^{-4}
	Number of function evaluations	18,027	16,874	12,674	10,985
f_{16}	Average best so-far	−1.0316	−1.0316	−1.0316	−1.0316
	Median best so-far	−1.0305	−1.0309	−1.0308	−1.0312
	Average mean fitness	−1.0307	−1.0306	−1.0298	−1.0312
	Number of function evaluations	19,857	17,024	12,387	10,087
f_{17}	Average best so-far	0.3979	0.3979	0.3979	0.3979
	Median best so-far	0.3932	0.3970	0.3972	0.3978
	Average mean fitness	0.3940	0.3877	0.3974	0.3978
	Number of function evaluations	18,952	16,725	11,697	9547
f_{18}	Average best so-far	−3.8696	−3.8646	−3.8696	−3.8697
	Median best so-far	−3.8642	−3.8597	−3.8621	−3.8690
	Average mean fitness	−3.8632	−3.8854	−3.8527	−3.8684
	Number of function evaluations	19,623	15,596	11,057	9874
f_{19}	Average best so-far	−3.7981	−3.5487	−3.7051	−3.8597
	Median best so-far	−3.2251	−3.3982	−3.5387	−3.7314
	Average mean fitness	−3.6571	−3.4247	−3.6257	−3.8024
	Number of function evaluations	20,147	16,087	12,074	10,087
f_{20}	Average best so-far	−3.3100	−3.3010	−3.3070	−3.3198
	Median best so-far	−3.3021	−3.2178	−3.3000	−3.3150
	Average mean fitness	−3.3076	−3.1871	−3.3047	−3.3171
	Number of function evaluations	22,387	17,287	11,854	9587
f_{21}	Average best so-far	−9.3541	−7.8514	−8.9862	−10.1532
	Median best so-far	−8.7431	−6.7481	−7.6217	−10.1008
	Average mean fitness	−9.2588	−7.0047	−8.0024	−10.1311
	Number of function evaluations	20,547	18,691	12,874	9675

(continued)

Table 7.11 (continued)

		CSA-NN	PSO-RBF	ES-K	APRE
f_{22}	Average best so-far	−9.9852	−9.8974	−9.1074	−10.4030
	Median best so-far	−8.9874	−8.8437	−9.0241	−10.3800
	Average mean fitness	−8.8731	−8.7450	−8.7091	−10.3011
	Number of function evaluations	21,007	17,964	13,207	10,254
f_{23}	Average best so-far	−8.3581	−9.8861	−8.0021	−10.5078
	Median best so-far	−8.0549	−9.1941	−7.8247	−10.4821
	Average mean fitness	−8.1841	−9.0040	−7.9867	−10.5004
	Number of function evaluations	20,847	18,217	14,781	9011

Maximum number of iterations = 500

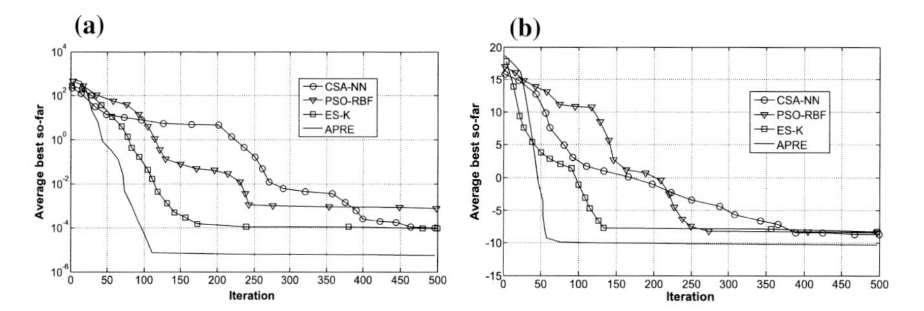

Fig. 7.7 Performance comparison of CSA-NN, PSO-RBF, ES-K and APRE for minimization of **a** f_{15} and **b** f_{22}

7.4 Summary

In this chapter, the performance of APRE as a global optimization algorithm has been presented. APRE reduces the number of function evaluations through the use of two mechanisms: (1) adapting dynamically the size of the population and (2) incorporating a fitness calculation strategy which decides when it is feasible to calculate or only estimate new generated individuals.

The algorithm begins with an initial population which will be used as a memory during the evolution process. To each memory element, it is assigned a normalized fitness value called quality factor that indicates the solution capacity provided by the element. From the memory, only a variable sub-set of elements is considered to be evolved. APRE generates new individuals considering two operators: seeking and refinement. Such operations are applied to improve the quality of the solutions by: (1) searching the unexplored solution space to identify promising areas containing better solutions than those found so far, and (2) successive refinement of the best found solutions. Once the new individuals are generated, the memory is updated. In such stage, the new individuals compete against the memory elements to build the final memory configuration.

In order to save computational time, the approach incorporates a fitness estimation strategy that decides which individuals can be estimated or actually evaluated. As a result, the approach can substantially reduce the number of function evaluations, yet preserving its good search capabilities. The APRE fitness calculation strategy estimates the fitness value of new individuals using memory elements located in neighboring positions that have been visited during the evolution process. In the strategy, those new individuals, close to a memory element whose quality factor is high, have a great probability to be evaluated by using the true objective function. Similarly, it is also evaluated those new particles lying in regions of the search space with no previous evaluations. The remaining search positions are estimated assigning them the same fitness value that the nearest location of the memory element. By the use of such fitness estimation method, the fitness value of only very few individuals are actually evaluated whereas the rest is just estimated.

Different to other approaches that use an already existent EA as framework, the APRE method has been completely designed to substantially reduce the computational cost, but preserving good search effectiveness.

APRE has been experimentally tested considering a challenging test suite gathering 23 benchmark functions. In order to analyze the performance of the APRE algorithm, it has been compared with other approaches that have been conceived to reduce the number of function evaluations. Conducted experiments, statistically validated, have demonstrated that the presented method achieves the best balance over its counterparts, in terms of both the number of function evaluations and solution accuracy.

References

1. Hamzaçebi, C.: Improving genetic algorithm performance by local search for continuous function optimization. Appl. Math. Comput. **196**(1), 309–317 (2008)
2. Kennedy, J., Eberhart, R.C.: Particle swarm optimization. In: Proceedings of the 1995 IEEE International Conference on Neural Networks, vol. 4, pp. 1942–1948 (1995)
3. de Castro, L.N., von Zuben, F.J.: Learning and optimization using the clonal selection principle. IEEE Trans. Evol. Comput. **6**(3), 239–251 (2002)
4. Schwefel, H.P.: Evolution and Optimum Seeking. Wiley, New York (1995)
5. Chafekar, D., Shi, L., Rasheed, K., Xuan, J.: Multi-objective GA optimization using reduced models. IEEE Trans. Syst. Man Cybern Part C **9**(2), 261–265 (2005)
6. Shan, S., Wang, G.G.: Survey of modeling and optimization strategies to solve high-dimensional design problems with computationally-expensive blackbox functions. Struct. Multi. Optim. **41**, 219–241 (2010)

7. Jin, Y.: Comprehensive survey of fitness approximation in evolutionary computation. Soft. Comput. **9**, 3–12 (2005)
8. Bernardino, H.S., Fonseca, L.G., Barbosa, H.J.C.: Surrogate-assisted clonal selection algorithms for expensive optimization problems. Evol. Intell. **4**, 81–97 (2011)
9. Regis, R.G.: Particle swarm with radial basis function surrogates for expensive black-box optimization. J. Comput. Sci. **5**, 12–23 (2014)
10. Tenne, Yoel: A computational intelligence algorithm for expensive engineering optimization problems. Eng. Appl. Artif. Intell. **25**, 1009–1021 (2012)
11. Jin, Y., Olhofer, M., Sendhoff, B.: A framework for evolutionary optimization with approximate fitness functions. IEEE Trans. Evol. Comput. **6**(5), 481–494 (2002)
12. Ong, Y.S., Nair, P.B., Keane, A.J.: Evolutionary optimization of computationally expensive problems via surrogate modeling. AIAA J. **41**(4), 687–696 (2003)
13. Chen, D., Zhao, C.: Particle swarm optimization with adaptive population size and its application. Appl. Soft Comput. **9**, 39–48 (2009)
14. Zhu, Wu, Tang, Y., Fang, J.-A., Zhang, W.: Adaptive population tuning scheme for differential evolution. Inf. Sci. **223**, 164–191 (2013)
15. Brest, J., Maučec, M.S.: Population size reduction for the differential evolution algorithm. Appl Intell. **29**, 228–247 (2008)
16. Cuevas, E., Santuario, E., Zaldívar, D., Perez-Cisneros, M.: Automatic circle detection on images based on an evolutionary algorithm that reduces the number of function evaluations. Math. Probl. Eng. **2013**, 17. Article ID 868434
17. Tan, K.C., Chiam, S.C., Mamun, A.A., Goh, C.K.: Balancing exploration and exploitation with adaptive variation for evolutionary multi-objective optimization. Eur. J. Oper. Res. **197**, 701–713 (2009)
18. Alba, E., Dorronsoro, B.: The exploration/exploitation tradeoff in dynamic cellular genetic algorithms. IEEE Trans. Evol. Comput. **9**(3), 126–142 (2005)
19. Ostadmohammadi, B., Mirzabeygi, P., Panahi, M.: An improved PSO algorithm with a territorial diversity-preserving scheme and enhanced exploration–exploitation balance. Swarm and Evolutionary Computation (in press)
20. Črepineš, M., Liu, S.H., Mernik, M.: Exploration and exploitation in evolutionary algorithms: A survey. ACM Comput. Surv. **1**(1), 1–33 (2011)
21. Storn, R., Price, K: Differential evolution-a simple and efficient adaptive scheme for global optimisation over continuous spaces. Tech. Rep. TR-95–012, ICSI, Berkeley, California (1995)
22. Fan, H.-Y., Lampinen, J.: A trigonometric mutation operation to differential evolution. J. Global Optim. **27**(1), 105–129 (2003)
23. Ali, M.M., Khompatraporn, C., Zabinsky, Z.B.: A numerical evaluation of several stochastic algorithms on selected continuous global optimization test problems. J. Global Optim. **31**(4), 635–672 (2005)
24. Chelouah, R., Siarry, P.: A continuous genetic algorithm designed for the global optimization of multimodal functions. J. Heuristics **6**(2), 191–213 (2000)
25. Herrera, F., Lozano, M., Sánchez, A.M.: A taxonomy for the crossover operator for real-coded genetic algorithms: an experimental study. Int. J. Intell. Syst. **18**(3), 309–338 (2003)
26. Laguna, M., Martí, R.: Experimental testing of advanced scatter search designs for global optimization of multimodal functions. J. Global Optim. **33**(2), 235–255 (2005)
27. Lozano, M., Herrera, F., Krasnogor, N., Molina, D.: Real-coded memetic algorithms with crossover hill-climbing. Evol. Comput. **12**(3), 273–302 (2004)
28. Moré, J.J., Garbow, B.S., Hillstrom, K.E.: Testing unconstrained optimization software. ACM Trans. Math. Soft. **7**(1), 17–41 (1981)
29. Ortiz-Boyer, D., Hervás-Martınez, C., García-Pedrajas, N.: CIXL2: A crossover operator for evolutionary algorithms based on population features. J. Artif. Intell. Res. **24**(1), 1–48 (2005)
30. Price, K., Storn, R.M., Lampinen, J.A.: Differential Evolution: A Practical Approach to Global Optimization. Springer, New York (2005)

31. Rahnamayan, S., Tizhoosh, H.R., Salama, M.M.A.: Opposition-based differential evolution. IEEE Trans. Evol. Comput. **12**(1), 64–79 (2008)
32. Whitley, D., Rana, D., Dzubera, J., Mathias, E.: Evaluating evolutionary algorithms. Artif. Intell. **85**(1–2), 245–276 (1996)
33. Yao, X., Liu, Y., Lin, G.: Evolutionary programming made faster. IEEE Trans. Evol. Comput. **3**(2), 82–102 (1999)
34. Li, Fei Kang Junjie, Ma, Zhenyue: Rosenbrock artificial bee colony algorithm for accurate global optimization of numerical functions. Inf. Sci. **181**, 3508–3531 (2011)
35. Wilcoxon, F.: Individual comparisons by ranking methods. Biometrics **1**, 80–83 (1945)
36. Garcia, S., Molina, D., Lozano, M., Herrera, F.: A study on the use of non-parametric tests for analyzing the evolutionary algorithms' behaviour: A case study on the CEC'2005 special session on real parameter optimization. J Heurist. (2008) doi:10.1007/s10732-008-9080-4
37. Shilane, D., Martikainen, J., Dudoit, S., Ovaska, S.: A general framework for statistical performance comparison of evolutionary computation algorithms. Inf. Sci. **178**, 2870–2879 (2008)

Chapter 8
Collective Animal Behavior Algorithm for Multimodal Optimization Functions

8.1 Introduction

A large number of real-world problems can be considered as multimodal function optimization subjects. An objective function may have several global optima, i.e. several points holding objective function values which are equal to the global optimum. Moreover, it may exhibit some other local optima points whose objective function values lay nearby a global optimum. Since the mathematical formulation of a real-world problem often produces a multimodal optimization issue, finding all global or even these local optima would provide to the decision makers multiple options to choose from [1].

Several methods have recently been proposed for solving the multimodal optimization problem. They can be divided into two main categories: deterministic and stochastic (metaheuristic) methods. When facing complex multimodal optimization problems, deterministic methods, such as gradient descent method, the quasi-Newton method and the Nelder-Mead's simplex method, may get easily trapped into the local optimum as a result of deficiently exploiting local information. They strongly depend on a priori information about the objective function, yielding few reliable results.

Metaheuristic algorithms have been developed combined rules and randomness mimicking several phenomena. These phenomena include evolutionary processes (e.g., the evolutionary algorithm proposed by Fogel et al. [2]; De Jong [3], and Koza [4] and the genetic algorithms (GAs) proposed by Holland [5] and Goldberg [6]), immunological systems (e.g., the artificial immune systems proposed by de Castro et al. [7]), physical processes (e.g., simulated annealing proposed by Kirkpatrick et al. [8], electromagnetism-like proposed by İlker et al. [9] and the gravitational search algorithm proposed by Rashedi et al. [10]) and the musical process of searching for a perfect state of harmony (proposed by Geem et al. [11]; Lee and Geem [12]; Geem [13] and Gao et al. [14]).

E. Cuevas et al., *Advances of Evolutionary Computation: Methods and Operators*, Studies in Computational Intelligence 629,
DOI 10.1007/978-3-319-28503-0_8

Traditional GA's perform well for locating a single optimum but fail to provide multiple solutions. Several methods have been introduced into the GA's scheme to achieve multimodal function optimization, such as sequential fitness sharing [15, 16], deterministic crowding [17], probabilistic crowding [18], clustering based niching [19], clearing procedure [20], species conserving genetic algorithm [21], and elitist-population strategies [22]. However, algorithms based on the GA's do not guarantee convergence to global optima because of their poor exploitation capability. GA's exhibit other drawbacks such as the premature convergence which results from the loss of diversity in the population and becomes a common problem when the search continues for several generations. Such drawbacks [23] prevent the GA's from practical interest for several applications.

Using a different metaphor, other researchers have employed Artificial Immune Systems (AIS) to solve the multimodal optimization problems. Some examples are the clonal selection algorithm [24] and the artificial immune network (AiNet) [25, 26]. Both approaches use some operators and structures which attempt to algorithmically mimic the natural immune system's behavior of human beings and animals.

On other hand, many studies have been inspired by animal behavior phenomena in order to develop optimization techniques such as the Particle swarm optimization (PSO) algorithm which models the social behavior of bird flocking or fish schooling [27]. In recent years, there have been several attempts to apply the PSO to multi-modal function optimization problems [28, 29]. However, the performance of such approaches presents several flaws when it is compared to the other multi-modal metaheuristic counterparts [26].

Recently, the concept of individual-organization [30, 31] has been widely used to understand collective behavior of animals. The central principle of individual-organization is that simple repeated interactions between individuals can produce complex behavioral patterns at group level [30, 32, 33]. Such inspiration comes from behavioral patterns seen in several animal groups, such as ant pheromone trail networks, aggregation of cockroaches and the migration of fish schools, which can be accurately described in terms of individuals following simple sets of rules [34]. Some examples of these rules [33, 35] include keeping current position (or location) for best individuals, local attraction or repulsion, random movements and competition for the space inside of a determined distance. On the other hand, new studies have also shown the existence of collective memory in animal groups [36–38]. The presence of such memory establishes that the previous history, of group structure, influences the collective behavior exhibited in future stages. Therefore, according to these new developments, it is possible to model complex collective behaviors by using simple individual rules and configuring a general memory.

This chapter proposes a new optimization algorithm inspired by the collective animal behavior. In this algorithm, the searcher agents are a group of animals that interact to each other based on simple behavioral rules which are modeled as mathematical operators. Such operations are applied to each agent considering that the complete group has a memory which stores its own best positions seen so far by applying a competition principle. The presented approach has also been compared to some other well-known metaheuristic search methods. The obtained results confirm a high performance of the presented method for solving various benchmark functions.

This chapter is organized as follows: Sect. 8.2 introduces the basic biologic aspects of the algorithm. In Sect. 8.3, the presented algorithm and its characteristics are described. A comparative study is presented in Sect. 8.4 and finally in Sect. 8.5 the conclusions are discussed.

8.2 Biologic Fundaments

The remarkable collective behavior of organisms such as swarming ants, schooling fish and flocking birds has long captivated the attention of naturalists and scientists. Despite a long history of scientific investigation, just recently we are beginning to decipher the relationship between individuals and group-level properties [39]. Grouping individuals often have to make rapid decisions about where to move or what behavior to perform, in uncertain and dangerous environments. However, each individual typically has only relatively local sensing ability [40]. Groups are, therefore, often composed of individuals that differ with respect to their informational status and individuals are usually not aware of the informational state of others [41], such as whether they are knowledgeable about a pertinent resource, or of a threat.

Animal groups are based on a hierarchic structure [42] which differentiates individuals according to a fitness principle known as Dominance [43]. Such concept represents the domain of some individuals within a group and occurs when competition for resources leads to confrontation. Several studies [44, 45] have found that such animal behavior lead to stable groups with better cohesion properties among individuals.

Recent studies have illustrated how repeated interactions among grouping animals scale to collective behavior. They have also remarkably revealed, that collective decision-making mechanisms across a wide range of animal group types, ranging from insects to birds (and even among humans in certain circumstances) seem to share similar functional characteristics [30, 34, 46]. Furthermore, at a certain level of description, collective decision-making in organisms shares essential common features such as a general memory. Although some differences may arise, there are good reasons to increase communication between researchers working in collective animal behavior and those involved in cognitive science [33].

Despite the variety of behaviors and motions of animal groups, it is possible that many of the different collective behavioral patterns are generated by simple rules followed by individual group members. Some authors have developed different models, such as the self-propelled particle (SPP) model which attempts to capture the collective behavior of animal groups in terms of interactions between group members following a diffusion process [47–50].

On other hand, following a biological approach, Couzin [33], Couzin and Krause [34] have proposed a model in which individual animals follow simple rules of thumb: (1) keep the position of best individuals; (2) move from or to nearby neighbors (local attraction or repulsion); (3) move randomly and (4) compete for the space inside of a determined distance. Each individual thus admits three different movements: attraction, repulsion or random, while holds two kinds of states: preserve the position or compete for a determined position. In the model, the movement experimented by each individual is decided randomly (according to an internal motivation), meanwhile the states are assumed according to a fixed criteria.

The dynamical spatial structure of an animal group can be explained in terms of its history [47]. Despite this, the majority of the studies have failed in considering the existence of memory in behavioral models. However, recent researches [36, 51] have also shown the existence of collective memory in animal groups. The presence of such memory establishes that the previous history of the group structure, influences the collective behavior exhibited in future stages. Such memory can contain the position of special group members (the dominant individuals) or the averaged movements produced by the group.

According to these new developments, it is possible to model complex collective behaviors by using simple individual rules and setting a general memory. In this work, the behavioral model of animal groups is employed for defining the evolutionary operators through the presented metaheuristic algorithm. A memory is incorporated to store best animal positions (best solutions) considering a competition-dominance mechanism.

8.3 Collective Animal Behaviour Algorithm (CAB)

The CAB algorithm assumes the existence of a set of operations that resembles the interaction rules that model the collective animal behavior. In the approach, each solution within the search space represents an animal position. The "fitness value" refers to the animal dominance with respect to the group. The complete process mimics the collective animal behavior.

The approach in this chapter implements a memory for storing best solutions (animal positions) mimicking the aforementioned biologic process. Such memory is divided into two different elements, one for maintaining the best found positions in each generation (M_g) and the other for storing best history positions during the complete evolutionary process ($\mathbf{M}_h$).

8.3.1 Description of the CAB Algorithm

Likewise other metaheuristic approaches, the CAB algorithm is also an iterative process. It starts by initializing the population randomly, i.e. generating random solutions or animal positions. The following four operations are thus applied until the termination criterion is met, i.e. the iteration number *NI* is reached as follows:

1. Keep the position of the best individuals.
2. Move from or nearby neighbors (local attraction and repulsion).
3. Move randomly.
4. Compete for the space inside of a determined distance (updating the memory).

8.3.1.1 Initializing the Population

The algorithm begins by initializing a set A of N_p animal positions $\left(\mathbf{A} = \{\mathbf{a}_1, \mathbf{a}_2, \ldots, \mathbf{a}_{Np}\}\right)$. Each animal position $\mathbf{a}_i$ is a *D*-dimensional vector containing parameter values to be optimized. Such values are randomly and uniformly distributed between the pre-specified lower initial parameter bound a_j^{low} and the upper initial parameter bound a_j^{high}.

$$a_{i,j} = a_j^{low} + \text{rand}(0,1) \cdot (\text{a}_j^{high} - \text{a}_j^{low}); \quad j = 1,2,\ldots,D; \quad i = 1,2,\ldots,N_p. \tag{8.1}$$

with *j* and *i* being the parameter and individual indexes respectively. Hence, $a_{i,j}$ is the *j*th parameter of the *i*th individual.

All the initial positions A are sorted according to the fitness function (dominance) to form a new individual set $\mathbf{X} = \{\mathbf{x}_1, \mathbf{x}_2, \ldots \mathbf{x}_{Np}\}$, so that we can choose the best *B* positions and store them in the memory $\mathbf{M}g$ and $\mathbf{M}_h$. The fact that both memories share the same information is only allowed at this initial stage.

8.3.1.2 Keep the Position of the Best Individuals

Analogously to the biological metaphor, this behavioral rule, typical in animal groups, is implemented as an evolutionary operation in our approach. In this operation, the first *B* elements of the new animal position set $\text{A}(\{\mathbf{a}_1, \mathbf{a}_2, \ldots, \mathbf{a}_B\})$ are generated. Such positions are computed by the values contained in the historic memory $\mathbf{M}_h$ considering a slight random perturbation around them. This operation can be modelled as follows:

$$\mathbf{a}_l = \mathbf{m}_h^l + \mathbf{v} \tag{8.2}$$

where $l \in \{1, 2, \ldots, B\}$ while $\mathbf{m}_h^l$ represents the *l*-element of the historic memory $\mathbf{M}_h$ and $\mathbf{v}$ is a random vector holding an appropriate small length.

8.3.1.3 Move from or to Nearby Neighbours

From the biological inspiration, where animals experiment a random local attraction or repulsion according to an internal motivation, we implement the evolutionary operators that mimic them. For this operation, a uniform random number r_m is generated within the range [0,1]. If r_m is less than a threshold H, a determined individual position is moved (attracted or repelled) considering the nearest best historical value of the group (the nearest position contained in $\mathbf{M}_h$) otherwise it is considered the nearest best value in the group of the current generation (the nearest position contained in $\mathbf{M}_g$). Therefore such operation can be modeled as follows:

$$\mathbf{a}_i \begin{cases} \mathbf{x}_i \pm r \cdot (\mathbf{m}_h^{nearest} - \mathbf{x}_i) & \text{with probability } H \\ \mathbf{x}_i \pm r \cdot (\mathbf{m}_g^{nearest} - \mathbf{x}_i) & \text{with probability } (1 - H) \end{cases} \quad (8.3)$$

where $i \in \{B+1, B+2, \ldots, N_p\}$, $\mathbf{m}_h^{nearest}$ and $\mathbf{m}_g^{nearest}$ represent the nearest elements of $\mathbf{M}_h$ and $\mathbf{M}_g$ to $\mathbf{x}_i$, while r is a random number between [−1,1]. Therefore, if $r > 0$, the individual position $\mathbf{x}_i$ is attracted to the position $\mathbf{m}_h^{nearest}$ or $\mathbf{m}_g^{nearest}$, otherwise such movement is considered as a repulsion.

8.3.1.4 Move Randomly

Following the biological model, under some probability P an animal randomly changes its position. Such behavioral rule is implemented considering the next expression:

$$\mathbf{a}_i = \begin{cases} \mathbf{r} & \text{with probability } P \\ \mathbf{x}_i & \text{with probability } (1 - P) \end{cases} \quad (8.4)$$

being $i \in \{B+1, B+2, \ldots, N_p\}$ and r a random vector defined within the search space. This operator is similar to re-initialize the particle in a random position as it is done by Eq. (8.1).

8.3.1.5 Compete for the Space Inside of a Determined Distance (Updating the Memory)

Once the operations to preserve the position of the best individuals, to move from or to nearby neighbors and to move randomly, have all been applied to the all N_p animal positions, generating N_p new positions, it is necessary to update the memory $\mathbf{M}_h$.

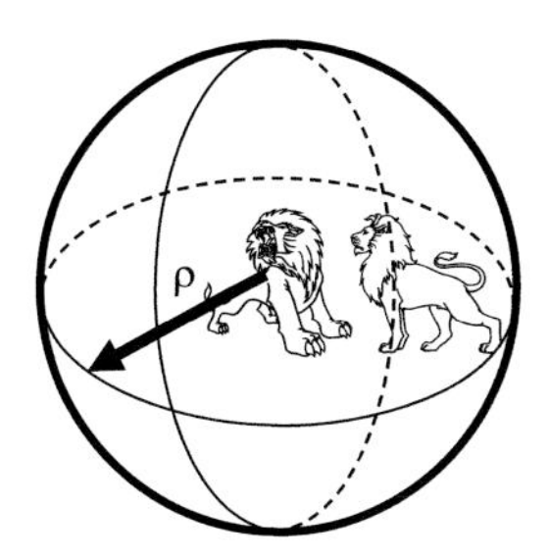

Fig. 8.1 Dominance concept, presented when two animals confront each other inside of a ρ distance

In order to update de memory $\mathbf{M}_h$, the concept of dominance is used. Animals that interact in a group keep a minimum distance among them. Such distance ρ depends on how aggressive the animal behaves [43, 51]. Hence, when two animals confront each other inside of such distance, the most dominant individual prevails as the other withdraws. Figure 8.1 shows this process.

In the presented algorithm, the historic memory $\mathbf{M}_h$ is updated considering the following procedure:

1. The elements of $\mathbf{M}_h$ and $\mathbf{M}_g$ are merged into $\mathbf{M}_U \cdot (\mathbf{M}_U = \mathbf{M}_h \cup \mathbf{M}_g)$.
2. Each element $\mathbf{m}_U^i$ of the memory $\mathbf{M}_U$, it is compared pair-wise with the remainder memory elements $(\{\mathbf{m}_U^1, \mathbf{m}_U^2 \ldots, \mathbf{m}_U^{2B-1}\}$. If the distance between both elements is less than ρ, the element holding a better performance in the fitness function will prevail meanwhile the other will be removed.
3. From the resulting elements of $\mathbf{M}_U$ (as they are obtained in step (2), the B best value is selected to integrate the new $\mathbf{M}_h$.

Unsuitable values of ρ result in a lower convergence rate, longer computation time, larger function evaluation number, convergence to a local maximum or unreliability of solutions. The ρ value is computed considering the following equation:

$$\rho = \frac{\prod_{j=1}^{D} (a_j^{high} - a_j^{low})}{10 \cdot D} \tag{8.5}$$

where a_j^{low} and a_j^{high} represent the pre-specified lower bound and the upper bound of the j-parameter respectively, within an D-dimensional space.

8.3.1.6 Computational Procedure

The computational procedure for the presented algorithm can be summarized as follows:

Step 1	Set the parameters N_p , B, H, P and NI
Step 2	Generate randomly the position set $\mathbf{A} = \{\mathbf{a}_1, \mathbf{a}_2, \ldots \mathbf{a}_{Np}\}$ using Eq. 8.1
Step 3	Sort A, according to the objective function (dominance), building $\mathbf{X} = \{\mathbf{x}_1, \mathbf{x}_2, \ldots, \mathbf{x}_{Np}\}$
Step 4	Choose the first B positions of X and store them into the memory $\mathbf{M}_g$
Step 5	Update $\mathbf{M}_h$ according to Sect. 8.3.1.5 (for the first iteration $\mathbf{M}_h = \mathbf{M}_g$)
Step 6	Generate the first B positions of the new solution set $\mathrm{A} = \{\mathbf{a}_1, \mathbf{a}_2, \ldots \mathbf{a}_B\}$. Such positions correspond to elements of $\mathbf{M}_h$ making a slight random perturbation around them. $\mathbf{a}_l = \mathbf{m}_h^l + \mathbf{v}$; being v a random vector holding an appropriate small length

Step 7	Generate the rest of the A elements using the attraction, repulsion and random movements.
	for i=B+1: N_p
	if ($r_1 < 1$-P) then
	attraction and repulsion movement
	{ if ($r_2 < H$) then
	$\mathbf{a}_i = \mathbf{x}_i \pm r \cdot (\mathbf{m}_h^{nearest} - \mathbf{x}_i)$
	else if
	$\mathbf{a}_i = \mathbf{x}_i \pm r \cdot (\mathbf{m}_g^{nearest} - \mathbf{x}_i)$
	}
	else if
	random movement
	{
	$\mathbf{a}_i = \mathbf{r}$
	}
	end for
	
	where $r_1, r_2, r \in \mathrm{rand}(0,1)$.
	where $\mathrm{r}_1, \mathrm{r}_2, \cdot\mathrm{rand}(0, 1)$

Step 8	If NI is completed, the process is thus completed; otherwise go back to step 3

8.3.1.7 Optima Determination

Just after the optimization process has finished, an analysis of the final $\mathbf{M}_h$ memory is executed in order to find the global and significant local minima. For it, a threshold valuw T_h is defined to decide which elements will be considered as a significant local minimum. Such threshold is thus computed as:

$$T_h = \frac{\max_{fitness}(\mathbf{M_h})}{6} \tag{8.6}$$

where $\max_{fitness}(\mathbf{M_h})$ represents the best fitness value among $\mathbf{M}_h$ elements. Therefore, memory elements whose fitness values are greater than T_h will be considered as global and local optima as other elements are discarded.

8.3.1.8 Numerical Example

In order to demonstrate the algorithm's step-by-step operation, a numerical example has been set by applying the presented method to optimize a simple function which is defined as follows:

$$\begin{aligned} f(x_1, x_2) = & e^{-((x_1-4)^2-(x_2-4)^2)} + e^{-((x_1+4)^2-(x_2-4)^2)} + 2 \cdot e^{-((x_1)^2+(x_2)^2)} \\ & + 2 \cdot e^{-((x_1)^2-(x_2+4)^2)} \end{aligned} \tag{8.7}$$

Considering the interval of $-5 \leq x_1, x_2, \leq 5$, the function possesses two global maxima of value 2 at $(x_1, x_2) = (0, 0)$ and. Likewise, it holds two local minima of value 1 at (−4,4) and (4,4). Figure 8.2a shows the 3D plot of this function. The parameters for the CAB algorithm are set as: $N_p = 10$, $B = 4$, $\mathrm{H} = 0.8$, $P = 0.1$, $\rho = 0$ and $NI = 30$.

Like all evolutionary approaches, CAB is a population-based optimizer that attacks the starting point problem by sampling the objective function at multiple, randomly chosen, initial points. Therefore, after setting parameter bounds that define the problem domain, 10 (N_p) individuals ($\mathbf{i}_1$, $\mathbf{i}_2$, … , $\mathbf{i}_{10}$) are generated using Eq. 8.1. Following an evaluation of each individual through the objective function (Eq. 8.5), all are sorted decreasingly in order to build vector $\mathbf{X} = (\mathbf{x}_1, \mathbf{x}_2, \ldots, \mathbf{x}_{10})$. Fig. 8.2b depicts the initial individual distribution in the search space. Then, both memories $\mathbf{M}_g\left(\mathbf{m}_g^1, \ldots, \mathbf{m}_g^4\right)$ and $\mathbf{M}_h\left(\mathbf{m}_h^1, \ldots, \mathbf{m}_h^4\right)$ are filled with the first four (B) elements present in X. Such memory elements are represented by solid points in Fig. 8.2c.

The new 10 individuals ($\mathbf{a}_1$, $\mathbf{a}_2$, … , $\mathbf{a}_{10}$) are evolved at each iteration following three different steps: 1. Keep the position of best individuals. 2. Move from or nearby neighbors and 3. Move randomly. The first new four elements ($\mathbf{a}_1$, $\mathbf{a}_2$, $\mathbf{a}_3$, $\mathbf{a}_4$) are generated considering the first step (Keeping the position of best individuals). Following such step, new individual positions are calculated as perturbed versions of all the elements which are contained in the $\mathbf{M}_h$ memory (that represent the best

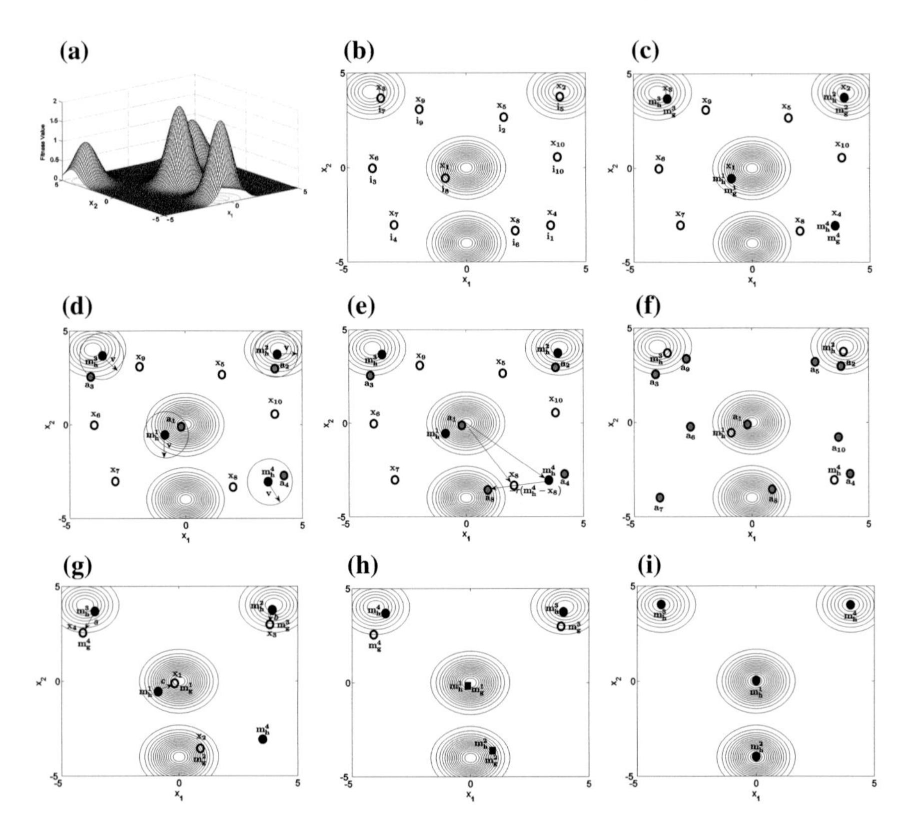

Fig. 8.2 CAB numerical example: **a** 3D plot of the function used as example. **b** Initial individual distribution. **c** Initial configuration of memories $\mathbf{M}_g$ and $\mathbf{M}_h$. **d** The computation of the first four individuals ($\mathbf{a}_1$, $\mathbf{a}_2$, $\mathbf{a}_3$, $\mathbf{a}_4$). **e** It shows the procedure employed by step 2 in order to calculate the new individual position $\mathbf{a}_8$. **f** Positions of all new individuals ($\mathbf{a}_1$, $\mathbf{a}_2$, … , $\mathbf{a}_{10}$). **g** Application of the dominance concept over elements of $\mathbf{M}_g$ and $\mathbf{M}_h$. **h** Final memory configurations of $\mathbf{M}_g$ and $\mathbf{M}_h$ after the first iteration. **i** Final memory configuration of $\mathbf{M}_h$ after 30 iterations

individuals known so far). Such perturbation is done by using $\mathbf{a}_l = \mathbf{m}_h^l + \mathbf{v} \cdot (l \in 1,\ldots 4)$. Figure 8.2d shows a comparative view between the memory element positions and the perturbed values of ($\mathbf{a}_1$, $\mathbf{a}_2$, $\mathbf{a}_3$, $\mathbf{a}_4$).

The remaining 6 new positions ($\mathbf{a}_5$, … , $\mathbf{a}_{10}$) are individually computed according to step 2 and 3. For such operation, a uniform random number r_1 is generated within the range [0, 1]. If r_1 is less than $1 - P$, the new position $\mathbf{a}_j$ ($j \in 5, \ldots, 10$) is generated through step 2; otherwise, $\mathbf{a}_j$ is obtained from a random re-initialization (step 3) between search bounds. In order to calculate a new position $\mathbf{a}_j$ at step 2, a decision must be made on whether it should be generated by using the elements of $\mathbf{M}_h$ or $\mathbf{M}_g$. For such decision, a uniform random number r_2 is generated within the range [0, 1]. If r_2 is less than H, the new position $\mathbf{a}_j$ is generated by using $\mathbf{x}_j \pm \mathrm{r} \cdot \left(\mathbf{m}_h^{nearets} - \mathbf{x}_j\right)$; otherwise, $\mathbf{a}_j$ is obtained by considering $\mathbf{x}_j \pm \mathrm{r} \cdot \left(\mathbf{m}_g^{nearest} - \mathbf{x}_j\right)$. Where $\mathbf{m}_h^{nearest}$ and

$\mathbf{m}_g^{nearest}$ represent the closest elements to $\mathbf{x}_j$ in memory $\mathbf{M}_h$ and $\mathbf{M}_g$ respectively. In the first iteration, since there is not available information from previous steps, both memories $\mathbf{M}_h$ and $\mathbf{M}_g$ share the same information which is only allowed at this initial stage. Figure 8.2e shows graphically the whole procedure employed by step 2 in order to calculate the new individual position $\mathbf{a}_8$ whereas Fig. 8.2f presents the positions of all new individuals ($\mathbf{a}_1$, $\mathbf{a}_2$, … , $\mathbf{a}_{10}$).

Finally, after all new positions ($\mathbf{a}_1$, $\mathbf{a}_2$, … , $\mathbf{a}_{10}$) have been calculated, memories $\mathbf{M}_h$ and $\mathbf{M}_g$ must be updated. In order to update $\mathbf{M}_h$, new calculated positions ($\mathbf{a}_1$, $\mathbf{a}_2$, … , $\mathbf{a}_{10}$) are arranged according to their fitness values by building vector $\mathbf{X} = (\mathbf{x}_1, \mathbf{x}_2, \ldots, \mathbf{x}_{10})$. Then, the elements of $\mathbf{M}_h$ are replaced by the first four elements in X (the best individuals of its generation). In order to calculate the new elements of $\mathbf{M}_{h,}$ current elements of $\mathbf{M}_h$ (the present values) and $\mathbf{M}_g$ (the updated values) are merged into $\mathbf{M}_U$. Then, by using the dominance concept (explained in Sect. 8.3.1.5) over $\mathbf{M}_U$, the best four values are selected to replace the elements in $\mathbf{M}_g$. Figure 8.2g, h show the updating procedure for both memories. Applying the dominance (see Fig. 8.2g), since the distances $a = dist\left(\mathbf{m}_h^3, \mathbf{m}_g^4\right), b = dist\left(\mathbf{m}_h^2, \mathbf{m}_g^2\right)$ and $c = dist\left(\mathbf{m}_h^1, \mathbf{m}_g^1\right)$ are less than $\rho = 3$, elements with better fitness evaluation will build the new memory $\mathbf{M}_h$. Figure 8.2h depicts final memory configurations. The circles and solid circles points represent the elements of $\mathbf{M}_g$ and $\mathbf{M}_h$ respectively whereas the bold squares perform as elements shared by both memories. Therefore, if the complete procedure is repeated over 30 iterations, the memory $\mathbf{M}_h$ will contain the 4 global and local maxima as elements. Figure 8.2i depicts the final configuration after 30 iterations.

8.4 Experimental Results

In this section, the performance of the presented algorithm is tested. Section 8.4.1 describes the experiment methodology. Sections 8.4.2, and 8.4.3 report on a comparison between the CAB experimental results and other multimodal metaheuristic algorithms for different kinds of optimization problems.

8.4.1 Experiment Methodology

In this section, we will examine the search performance of the proposed CAB by using a test suite of 8 benchmark functions with different complexities. They are listed in Tables 8.1 and 8.2. The suite mainly contains some representative, complicated and multimodal functions with several local optima. These functions are normally regarded as difficult to be optimized as they are particularly challenging to the applicability and efficiency of multimodal metaheuristic algorithms. The performance measurements considered at each experiment are the following:

Table 8.1 The test suite of multimodal functions for Experiment 8.4.2

Function	Search space	Sketch
$f_1 = \sin^6(5\pi x)$ Deb's function 5 optima	$x \in [0, 1]$	
$f_2(x) = 2^{-2((x-0.1)/0.9)^2} \cdot \sin(5\pi x)$ Deb's decreasing function 5 optima	$x \in [0, 1]$	
$f_3(z) = \frac{1}{1+\|z^6+1\|}$ Roots function 6 optima	$z \in C,\ z = x_1 + ix_2$ $x_1, x_2 \in [-2, 2]$	
$f_4(x_1, x_2) = x_1 \sin(4\pi x_1)\ldots$ $\ldots - x_2 \sin(4\pi x_2 + \pi) + 1$ Two dimensional multi-modal function 100 optima	$x_1, x_2 \in [-2, 2]$	

- The consistency of locating all known optima; and
- The averaged number of objective function evaluations that are required to find such optima (or the running time under the same condition).

The experiments compare the performance of CAB against the Deterministic Crowding [17], the Probabilistic Crowding [18], the Sequential Fitness Sharing [15], the Clearing Procedure [20], the Clustering Based Niching (CBN) [19], the Species Conserving Genetic Algorithm (SCGA) [21], the Elitist-population strategy (AEGA) [22], the Clonal Selection algorithm [24] and the artificial immune network (AiNet) [25].

Since the approach solves real-valued multimodal functions, we have used, in the GA-approaches, consistent real coding variable representation, uniform crossover and mutation operators for each algorithm seeking a fair comparison. The crossover probability Pc = 0.8 and the mutation probability Pm = 0.1 have been used. We use the standard tournament selection operator with a tournament size = 2 in our implementation of Sequential Fitness Sharing, Clearing Procedure, CBN, Clonal Selection algorithm, and SCGA. On the other hand, the parameter values for the aiNet algorithm have been defined as suggested in [25], with the mutation strength $\beta = 100$, the suppression threshold $\sigma_{s(aiNet)} = 0.2$ and the update rate $d = 40\ \%$.

Table 8.2 The test suite of multimodal functions used in the Experiment 8.4.3

Function	Search space	Sketch
$f_5(x_1, x_2) = -(20 + x_1^2 + x_2^2 - 10(\cos(2\pi x_1)\ldots$ $\ldots + \cos(2\pi x_2)))$ Rastringin's function 100 optima	$x_1, x_2 \in [-10, 10]$	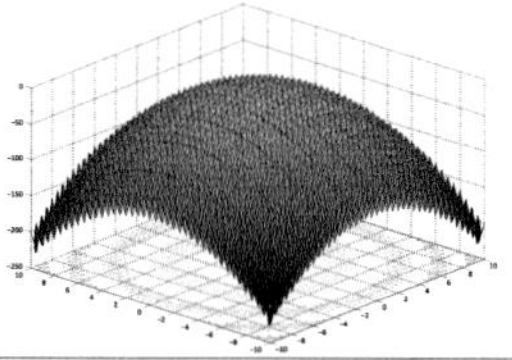
$f_6(x_1, x_2) = -\Pi_{i=1}^{2}\sum_{j=1}^{5}\cos((j+1)x_i + j)$ Shubert function 18 optima	$x_1, x_2 \in [-10, 10]$	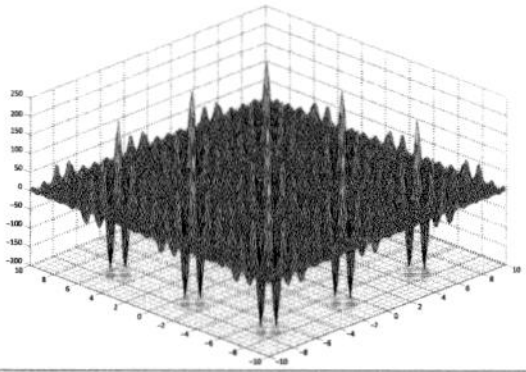
$f_7(x_1, x_2) = \frac{1}{4000}\sum_{i=1}^{2} x_i^2 - \prod_{i=1}^{2}\cos\left(\frac{x_i}{\sqrt{2}}\right) + 1$ Griewank function 100 optima	$x_1, x_2 \in [-100, 100]$	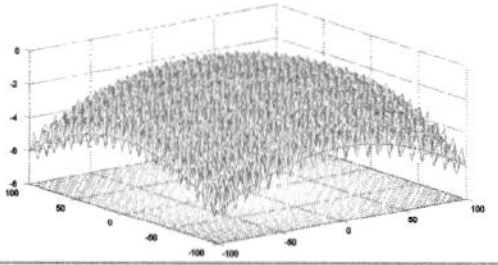
$f_8(x_1, x_2) = \frac{\cos(0.5x_1) + \cos(0.5x_2)}{4000}\ldots$ $\ldots + \cos(10x_1)\cos(10x_2)$ Modified Griewank function 100 optima	$x_1, x_2 \in [0, 120]$	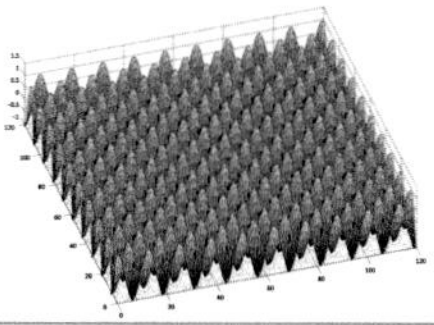

In the case of the CAB algorithm, the parameters are set to $N_p = 200, B = 100, P = 0.8$ and $H = 0.6$. Once they have been all experimentally determined, they are kept for all the test functions through all experiments.

To avoid relating the optimization results to the choice of a particular initial population and to conduct fair comparisons, we perform each test 50 times, starting from various randomly selected points in the search domain as it is commonly given in the literature. An optimum o_j is considered as found if $\exists x_i \cdot Pop(k = T)|d(x_i, o_j) < 0.005$, where $Pop(k = T)$ is the complete population at the end of the run T and x_i is an individual in $Pop(k = T)$.

All algorithms have been tested in MatLAB© over the same Dell Optiplex GX260 computer with a Pentium-4 2.66G-HZ processor, running Windows XP operating system over 1 Gb of memory. Next sections present experimental results for multimodal optimization problems which have been divided into two groups with different purposes. The first one consists of functions with smooth landscapes and well defined optima (local and global values), while the second gathers functions holding rough landscapes and complex location optima.

8.4.2 Comparing CAB Performance for Smooth Landscapes Functions

This section presents a performance comparison for different algorithms solving multimodal problems $f_1 - f_4$ in Table 8.1. The aim is to determine whether CAB is more efficient and effective than other existing algorithms for finding all multiple optima of $f_1 - f_4$. The stopping criterion analyzes if the number identified optima cannot be further increased over 10 successive generations after the first 100 generations, then the execution will be stopped. Four measurements have been employed to evaluate the performance:

- The average of optima found within the final population (NO);
- The average distance between multiple optima detected by the algorithm and their closest individuals in the final population (DO);
- The average of function evaluations (FE); and
- The average of execution time in seconds (ET).

Table 8.3 provides a summarized performance comparison among several algorithms. Best results have been bold-faced. From the NO measure, CAB always finds better or equally optimal solutions for the multimodal problems $f_1 - f_4$. It is evident that each algorithm can find all optima of f_1. For function f_2, only AEGA, Clonal Selection algorithm, aiNet, and CAB can eventually find all optima each time. For function f_3, Clearing Procedure, SCGA, AEGA and CAB can get all optima at each run. For function f_4, Deterministic Crowding leads to premature convergence and all other algorithms cannot get any better results but CAB yet can find all multiple optima 48 times in 50 runs and its average successful rate for each run is higher than 99 %. By analyzing the DO measure in Table 8.3, CAB has obtained the best score for all the multimodal problems except for f_3. In the case of f_3, the solution precision of CAB is only worse than that of Clearing Procedure. On the other hand, CAB has smaller standard deviations in the NO and DO measures than all other algorithms and hence its solution is more stable.

From the FE measure in Table 8.3, it is clear that CAB needs fewer function evaluations than other algorithms considering the same termination criterion. Recall that all algorithms use the same conventional crossover and mutation operators. It can be easily deduced from results that the CAB algorithm is able to produce better search positions (better compromise between exploration and exploitation), in a more efficient and effective way than other multimodal search strategies.

To validate that CAB improvement over other algorithms as a result of CAB producing better search positions over iterations, Fig. 8.3 shows the comparison of CAB and other multimodal algorithms for f_4. The initial populations for all algorithms have 200 individuals. In the final population of CAB, the 100 individuals belonging to the $\mathbf{M}_h$ memory correspond to the 100 multiple optima, while, on the contrary, the final population of the other nine algorithms fail consistently in finding all optima, despite they have superimposed several times over some previously found optima.

Table 8.3 Performance comparison among the multimodal optimization algorithms for the test functions $f_1 - f_4$

Function	Algorithm	NO	DO	FE	ET
f_1	Deterministic crowding	5(0)	$1.52 \times 10^{-4}(1.38 \times 10^{-4})$	7153 (358)	0.091(0.013)
	Probabilistic crowding	5(0)	$3.63 \times 10^{-4}(6.45 \times 10^{-5})$	10,304(487)	0.163(0.011)
	Sequential fitness sharing	5(0)	$4.76 \times 10^{-4}(6.82 \times 10^{-5})$	9927(691)	0.166(0.028)
	Clearing procedure	5(0)	$1.27 \times 10^{-4}(2.13 \times 10^{-5})$	5860(623)	0.128(0.021)
	CBN	5(0)	$2.94 \times 10^{-4}(4.21 \times 10^{-5})$	10,781(527)	0.237(0.019)
	SCGA	5(0)	$1.16 \times 10^{-4}(3.11 \times 10^{-5})$	6792(352)	0.131(0.009)
	AEGA	5(0)	$4.6 \times 10^{-4}(1.35 \times 10^{-5})$	2591(278)	0.039(0.007)
	Clonal selection algorithm	5(0)	$1.99 \times 10^{-4}(8.25 \times 10^{-5})$	15,803(381)	0.359(0.015)
	AiNet	5(0)	$1.28 \times 10^{-4}(3.88 \times 10^{-5})$	12,369(429)	0.421(0.021)
	CAB	5(0)	$\mathbf{1.69 \times 10^{-5}(5.2 \times 10^{-6})}$	1776(125)	0.020(0.009)
f_2	Deterministic crowding	3.53(0.73)	$3.61 \times 10^{-3}(6.88 \times 10^{-4})$	6026 (832)	0.271(0.06)
	Probabilistic crowding	4.73(0.64)	$2.82 \times 10^{-3}(8.52 \times 10^{-4})$	10,940(9517)	0.392(0.07)
	Sequential fitness sharing	4.77(0.57)	$2.33 \times 10^{-3}(4.36 \times 10^{-4})$	12,796(1430)	0.473(0.11)
	Clearing procedure	4.73(0.58)	$4.21 \times 10^{-3}(1.24 \times 10^{-3})$	8465(773)	0.326(0.05)
	CBN	4.70(0.53)	$2.19 \times 10^{-3}(4.53 \times 10^{-4})$	14,120(2187)	0.581(0.14)
	SCGA	4.83(0.38)	$3.15 \times 10^{-3}(4.71 \times 10^{-4})$	10,548(1382)	0.374(0.09)
	AEGA	5(0)	$1.38 \times 10^{-4}(2.32 \times 10^{-5})$	3605(426)	0.102(0.04)
	Clonal selection algorithm	5(0)	$1.37 \times 10^{-3}(6.87 \times 10^{-4})$	21,922(746)	0.728(0.06)
	AiNet	5(0)	$1.22 \times 10^{-3}(5.12 \times 10^{-4})$	18,251(829)	0.664(0.08)
	CAB	5(0)	$\mathbf{4.5 \times 10^{-5}(8.56 \times 10^{-6})}$	2065(92)	0.08(0.007)
f_3	Deterministic crowding	4.23(1.17)	$7.79 \times 10^{-4}(4.76 \times 10^{-4})$	11,009 (1137)	1.07(0.13)
	Probabilistic crowding	4.97(0.64)	$2.35 \times 10^{-3}(7.14 \times 10^{-4})$	16,391(1204)	1.72(0.12)
	Sequential fitness sharing	4.87(0.57)	$2.56 \times 10^{-3}(2.58 \times 10^{-3})$	14,424(2045)	1.84(0.26)
	Clearing procedure	6(0)	$\mathbf{7.43 \times 10^{-5}(4.07 \times 10^{-5})}$	12,684(1729)	1.59(0.19)
	CBN	4.73(1.14)	$1.85 \times 10^{-3}(5.42 \times 10^{-4})$	18,755(2404)	2.03(0.31)
	SCGA	6(0)	$3.27 \times 10^{-4}(7.46 \times 10^{-5})$	13,814(1382)	1.75(0.21)
	AEGA	6(0)	$1.21 \times 10^{-4}(8.63 \times 10^{-5})$	6218(935)	0.53(0.07)
	Clonal selection algorithm	5.50(0.51)	$4.95 \times 10^{-3}(1.39 \times 10^{-3})$	25,953(2918)	2.55(0.33)
	AiNet	4.8(0.33)	$3.89 \times 10^{-3}(4.11 \times 10^{-4})$	20,335(1022)	2.15(0.10)
	CAB	6(0)	$9.87 \times 10^{-5}(1.69 \times 10^{-5})$	4359(75)	0.11(0.023)

(continued)

Table 8.3 (continued)

Function	Algorithm	NO	DO	FE	ET
f_4	Deterministic crowding	76.3(11.4)	$4.52 \times 10^{-3}(4.17 \times 10^{-3})$	1861,707(329,254)	21.63(2.01)
	Probabilistic crowding	92.8(3.46)	$3.46 \times 10^{-3}(9.75 \times 10^{-4})$	2,638,581(597,658)	31.24(5.32)
	Sequential fitness sharing	89.9(5.19)	$2.75 \times 10^{-3}(6.89 \times 10^{-4})$	2,498,257(374,804)	28.47(3.51)
	Clearing procedure	89.5(5.61)	$3.83 \times 10^{-3}(9.22 \times 10^{-4})$	2,257,964(742,569)	25.31(6.24)
	CBN	90.8(6.50)	$4.26 \times 10^{-3}(1.14 \times 10^{-3})$	2,978,385(872,050)	35.27(8.41)
	SCGA	91.4(3.04)	$3.73 \times 10^{-3}(2.29 \times 10^{-3})$	2,845,789(432,117)	32.15(4.85)
	AEGA	95.8(1.64)	$1.44 \times 10^{-4}(2.82 \times 10^{-5})$	1,202,318(784,114)	12.17(2.29)
	Clonal selection algorithm	92.1(4.63)	$4.08 \times 10^{-3}(8.25 \times 10^{-3})$	3,752,136(191,849)	45.95(1.56)
	AiNet	93.2(7.12)	$3.74 \times 10^{-3}(5.41 \times 10^{-4})$	2,745,967(328,176)	38.18(3.77)
	CAB	100(2)	$\mathbf{2.31 \times 10^{-5}(5.87 \times 10^{-6})}$	697,578(57,089)	5.78(1.26)

The standard unit in the column ET is seconds. (For all the parameters, numbers in parentheses are the standard deviations.) Bold-cased letters represents best obtained results

When comparing the execution time (ET) in Table 8.3, CAB uses significantly less time to finish than other algorithms. The situation can be registered by the reduction of the redundancy in the $\mathbf{M}_h$ memory due to competition (dominance) criterion. All these comparisons show that CAB generally outperforms all other multimodal algorithms regarding efficacy and efficiency.

8.4.3 Comparing CAB Performance in Rough Landscapes Functions

This section presents the performance comparison among different algorithms solving multimodal optimization problems which are listed in Table 8.2. Such problems hold lots of local optima and very rugged landscapes. The goal of multimodal optimizers is to find as many as possible global optima and possibly good local optima. Rastrigin's function f_5 and Griewank's function f_7 have 1 and 18 global optima respectively, becoming practical as to test to whether a multimodal algorithm can find a global optimum and at least 80 higher fitness local optima to validate the algorithms' performance.

Our main objective in these experiments is to determine whether CAB is more efficient and effective than other existing algorithms for finding the multiple high fitness optima of functions $f_5 - f_8$. In the experiments, the initial population size for all algorithms has been set to 1000. For Sequential Fitness Sharing, Clearing Procedure, CBN, Clonal Selection, SCGA, and AEGA, we have set the distance threshold σ_s to 5. The algorithms' stopping criterion checks whenever the number of optima found cannot be further increased in 50 successive generations after the

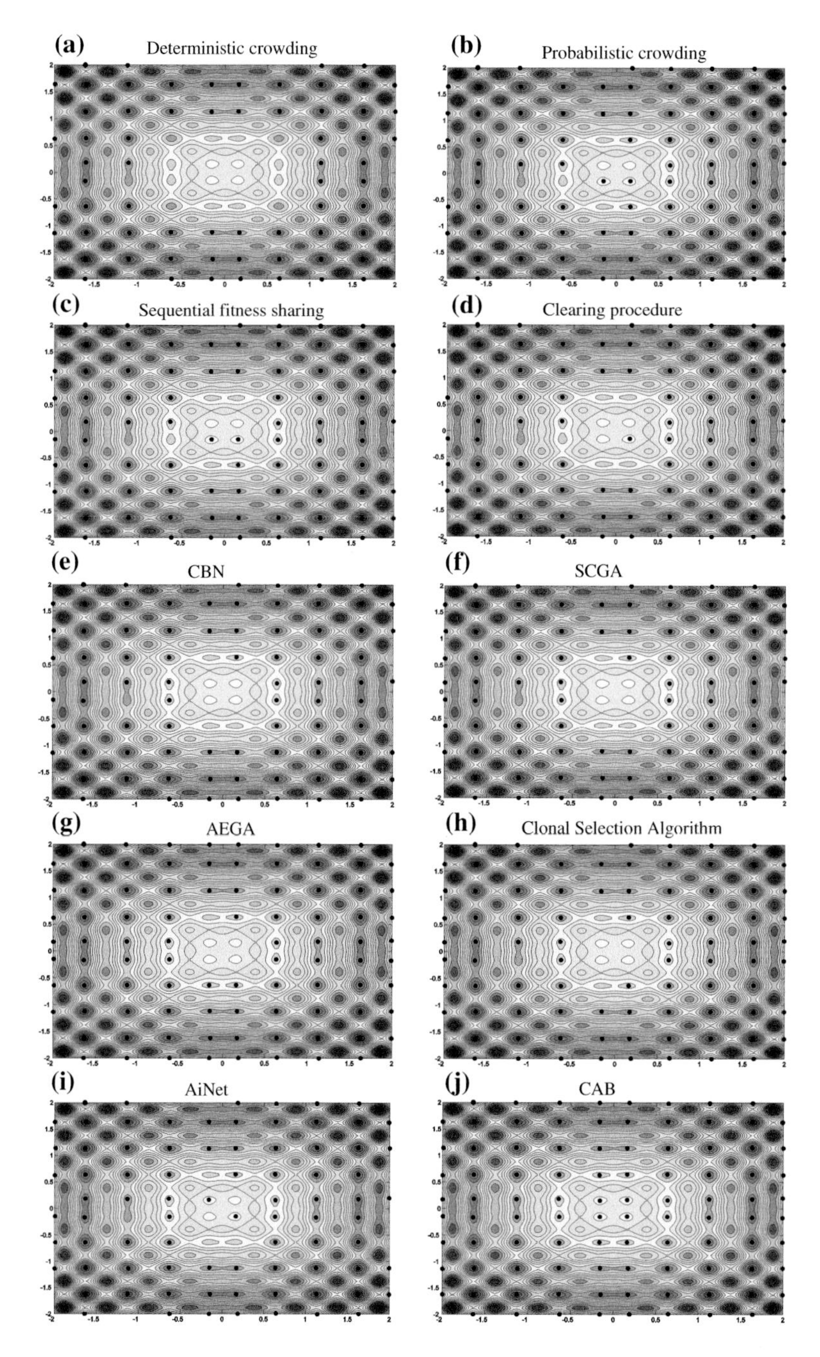

Fig. 8.3 Typical results of the maximization of f_4. **a–j** Local and global optima located by all ten algorithms in the performance comparison

Table 8.4 Performance comparison among multimodal optimization algorithms for the test functions $f_5 - f_8$

Function	Algorithm	NO	DO	FE	ET
f_5	Deterministic crowding	62.4(14.3)	$4.72 \times 10^{-3}(4.59 \times 10^{-3})$	1,760,199(254,341)	14.62(2.83)
	Probabilistic crowding	84.7(5.48)	$1.50 \times 10^{-3}(9.38 \times 10^{-4})$	2,631,627(443,522)	34.39(5.20)
	Sequential fitness sharing	76.3(7.08)	$3.51 \times 10^{-3}(1.66 \times 10^{-3})$	2,726,394(562,723)	36.55(7.13)
	Clearing procedure	93.6(2.31)	$2.78 \times 10^{-3}(1.20 \times 10^{-3})$	2,107,962(462,622)	28.61(6.47)
	CBN	87.9(7.78)	$4.33 \times 10^{-3}(2.82 \times 10^{-3})$	2,835,119(638,195)	37.05(8.23)
	SCGA	97.4(4.80)	$1.34 \times 10^{-3}(8.72 \times 10^{-4})$	2,518,301(643,129)	30.27(7.04)
	AEGA	99.4(1.39)	$6.77 \times 10^{-4}(3.18 \times 10^{-4})$	978,435(71,135)	10.56(4.81)
	Clonal selection algorithm	90.6(9.95)	$3.15 \times 10^{-3}(1.47 \times 10^{-3})$	5,075,208(194,376)	58.02(2.19)
	AiNet	93.8(7.8)	$2.11 \times 10^{-3}(3.2 \times 10^{-3})$	3,342,864(549,452)	51.65(6.91)
	CAB	100(2)	$2.22 \times \mathbf{10^{-4}}(3.1 \times \mathbf{10^{-5}})$	680,211(12,547)	7.33(1.84)
f_6	Deterministic crowding	9.37(1.91)	$3.26 \times 10^{-3}(5.34 \times 10^{-4})$	832,546(75,413)	4.58(0.57)
	Probabilistic crowding	15.17(2.43)	$2.87 \times 10^{-3}(5.98 \times 10^{-4})$	1,823,774(265,387)	12.92(2.01)
	Sequential fitness sharing	15.29(2.14)	$1.42 \times 10^{-3}(5.29 \times 10^{-4})$	1,767,562(528,317)	14.12(3.51)
	Clearing procedure	18(0)	$1.19 \times 10^{-3}(6.05 \times 10^{-4})$	1,875,729(265,173)	11.20(2.69)
	CBN	14.84(2.70)	$4.39 \times 10^{-3}(2.86 \times 10^{-3})$	2,049,225(465,098)	18.26(4.41)
	SCGA	4.83(0.38)	$1.58 \times 10^{-3}(4.12 \times 10^{-4})$	2,261,469(315,727)	13.71(1.84)
	AEGA	18(0)	$3.34 \times 10^{-4}(1.27 \times 10^{-4})$	656,639(84,213)	3.12(1.12)
	Clonal selection algorithm	18(0)	$3.42 \times 10^{-3}(1.58 \times 10^{-3})$	4,989,856(618,759)	33.85(5.36)
	AiNet	18(0)	$2.11 \times 10^{-3}(3.31 \times 10^{-3})$	3,012,435(332,561)	26.32(2.54)
	CAB	18(0)	$\mathbf{1.02 \times 10^{-4}(4.27 \times 10^{-5})}$	431,412(21,034)	2.21(0.51)
f_7	Deterministic crowding	52.6(8.86)	$3.71 \times 10^{-3}(1.54 \times 10^{-3})$	2,386,960(221,982)	19.10(2.26)
	Probabilistic crowding	79.2(4.94)	$3.48 \times 10^{-3}(3.79 \times 10^{-3})$	3,861,904(457,862)	43.53(4.38)
	Sequential fitness sharing	63.0(5.49)	$4.76 \times 10^{-3}(3.55 \times 10^{-3})$	3,619,057(565,392)	42.98(6.35)
	Clearing Procedure	79.4(4.31)	$2.95 \times 10^{-3}(1.64 \times 10^{-3})$	3,746,325(594,758)	45.42(7.64)
	CBN	71.3(9.26)	$3.29 \times 10^{-3}(4.11 \times 10^{-3})$	4,155,209(465,613)	48.23(5.42)
	SCGA	94.9(8.18)	$2.6310^{-3}(1.81 \times 10^{-3})$	3,629,461(373,382)	47.84(0.21)
	AEGA	98(2)	$1.31 \times 10^{-3}(8.76 \times 10^{-4})$	1,723,342(121,043)	12,54(1.31)
	Clonal selection algorithm	89.2(5.44)	$3.02 \times 10^{-3}(1.63 \times 10^{-3})$	5,423,739(231,004)	47.84(6.09)
	AiNet	92.7(3.21)	$2.79 \times 10^{-3}(3.19 \times 10^{-4})$	4,329,783(167,932)	41.64(2.65)
	CAB	100(1)	$\mathbf{3.32 \times 10^{-4}(5.25 \times 10^{-5})}$	953,832(9,345)	8.82(1.51)

(continued)

Table 8.4 (continued)

Function	Algorithm	NO	DO	FE	ET
f_8	Deterministic crowding	44.2(7.93)	$4.45 \times 10^{-3}(3.63 \times 10^{-3})$	2,843,452(353,529)	23.14(3.85)
	Probabilistic crowding	70.1(8.36)	$2.52 \times 10^{-3}(1.47 \times 10^{-3})$	4,325,469(574,368)	49.51(6.72)
	Sequential fitness sharing	58.2(9.48)	$4.14 \times 10^{-3}(3.31 \times 10^{-3})$	4,416,150(642,415)	54.43(12.6)
	Clearing procedure	67.5(10.11)	$2.31 \times 10^{-3}(1.43 \times 10^{-3})$	4,172,462(413,537)	52.39(7.21)
	CBN	53.1(7.58)	$4.36 \times 10^{-3}(3.53 \times 10^{-3})$	4,711,925(584,396)	61.07(8.14)
	SCGA	87.3(9.61)	$3.15 \times 10^{-3}(2.07 \times 10^{-3})$	3,964,491(432,117)	53.87(8.46)
	AEGA	90.6(1.65)	$2.55 \times 10^{-3}(9.55 \times 10^{-4})$	2,213,754(412,538)	16.21(3.19)
	Clonal selection algorithm	74.4(7.32)	$3.52 \times 10^{-3}(9.55 \times 10^{-4})$	5,835,452(498,033)	74.26(5.47)
	AiNet	83.2(6.23)	$3.11 \times 10^{-3}(2.41 \times 10^{-4})$	4,123,342(213,864)	60.38(5.21)
	CAB	97(2)	$\mathbf{1.54 \times 10^{-3}(4.51 \times 10^{-4})}$	1,121,523(51,732)	12.21(2.66)

The standard unit of the column ET is seconds (numbers in parentheses are standard deviations). Bold-case letters represent best results

first 500 generations. If such condition prevails then the algorithm is halted. We still evaluate the performance of all algorithms using the aforementioned four measures NO, DO, FE, and ET.

Table 8.4 provides a summary of the performance comparison among different algorithms. From the NO measure, we observe that CAB could always find more optimal solutions for the multimodal problems $f_5 - f_8$. For Rastrigin's function f_5, only CAB can find all multiple high fitness optima 49 times out of 50 runs and its average successful rate for each run is higher than 97 %. On the contrary, other algorithms cannot find all multiple higher fitness optima for any run. For f_6, 5 algorithms (Clearing Procedure, SCGA, AEGA, clonal selection algorithm, AiNet and CAB) can get all multiple higher fitness maxima for each run respectively. For Griewank's function (f_7), only CAB can get all multiple higher fitness optima for each run. In case of the modified Griewank's function (f_8), it has numerous optima whose value is always the same. However, CAB still can find all global optima with a effectiveness rate of 95 %.

From the FE and ET measures in Table 8.4, we can clearly observe that CAB uses significantly fewer function evaluations and a shorter running time than all other algorithms under the same termination criterion. Moreover, Deterministic Crowding leads to premature convergence as CAB is at least 2.5, 3.8, 4, 3.1, 4.1, 3.7, 1.4, 7.9 and 4.9 times faster than all others respectively according to Table 8.4 for functions $f_5 - f_8$.

8.5 Summary

In recent years, several metaheuristic optimization methods have been inspired from nature-like phenomena. In this article, a new multimodal optimization algorithm known as the Collective Animal Behavior Algorithm (CAB) has been introduced. In CAB, the searcher agents are a group of animals that interact to each other depending on simple behavioral rules which are modeled as mathematical operators. Such operations are applied to each agent considering that the complete group hold a memory to store its own best positions seen so far, using a competition principle.

CAB has been experimentally evaluated over a test suite consisting of 8 benchmark multimodal functions for optimization. The performance of CAB has been compared to some other existing algorithms including Deterministic Crowding [17], Probabilistic Crowding [18], Sequential Fitness Sharing [15], Clearing Procedure [20], Clustering Based Niching (CBN) [19], Species Conserving Genetic Algorithm (SCGA) [21], elitist-population strategies (AEGA) [22], Clonal Selection algorithm [24] and the artificial immune network (aiNet) [25]. All experiments have demonstrated that CAB generally outperforms all other multimodal metaheuristic algorithms regarding efficiency and solution quality, typically showing significant efficiency speedups. The remarkable performance of CAB is due to two different features: (i) operators allow a better exploration of the search space, increasing the capacity to find multiple optima; (ii) the diversity of solutions contained in the $\mathbf{M}_h$ memory in the context of multimodal optimization, is maintained and even improved through of the use of a competition principle (dominance concept).

References

1. Ahrari, A., Shariat-Panahi, M., Atai, A.A.: GEM: a novel evolutionary optimization method with improved neighbourhood search. Appl. Math. Comput. **210**(2), 376–386 (2009)
2. Fogel, L.J., Owens, A.J., Walsh, M.J.: Artificial Intelligence through simulated evolution. John Wiley, Chichester, UK (1966)
3. De Jong, K. Analysis of the behavior of a class of genetic adaptive systems. Ph.D. Thesis, University of Michigan, Ann Arbor, MI, 1975
4. Koza, J.R.: Genetic programming: a paradigm for genetically breeding populations of computer programs to solve problems, Rep. No. STAN-CS-90-1314. Stanford University, CA (1990)
5. Holland, J.H.: Adaptation in natural and artificial systems. University of Michigan Press, Ann Arbor, MI (1975)
6. Goldberg, D.E.: Genetic algorithms in search, optimization and machine learning. Addison Wesley, Boston, MA (1989)
7. de Castro, LN, Von Zuben, FJ. Artificial immune systems: Part I – basic theory and applications. Technical report, TR-DCA 01/99, 1999
8. Kirkpatrick, S., Gelatt, C., Vecchi, M.: Optimization by simulated annealing. Science **220** (4598), 671–680 (1983)

9. İlker, B., Birbil, S., Shu-Cherng, F.: An electromagnetism-like Mechanism for Global Optimization. J. Glob. Optim. **25**, 263–282 (2003)
10. Rashedi, E., Nezamabadi-Pour, H., Saryazdi, S.: GSA: a gravitational search algorithm. Inf. Sci. **179**, 2232–2248 (2009)
11. Geem, Z.W., Kim, J.H., Loganathan, G.V.: A new heuristic optimization algorithm: harmony search. Simulation **76**(2), 60–68 (2001)
12. Lee, K.S., Geem, Z.W.: A new meta-heuristic algorithm for continues engineering optimization: harmony search theory and practice. Comput. Meth. Appl. Mech. Eng. **194**, 3902–3933 (2004)
13. Geem, Z.W.: Novel derivative of harmony search algorithm for discrete design variables. Appl. Math. Comput. **199**, 223–230 (2008)
14. Gao, X.Z., Wang, X., Ovaska, S.J.: Uni-modal and multi-modal optimization using modified harmony search methods. Int. J. Innovative Comput. Inf. Control **5**(10(A)), 2985–2996 (2009)
15. Beasley, D., Bull, D.R., Matin, R.R.: A sequential niche technique for multimodal function optimization. Evol. Comput. **1**(2), 101–125 (1993)
16. Miller, B.L., Shaw, M.J. Genetic algorithms with dynamic niche sharing for multimodal function optimization. In: Proceedings of the 3rd IEEE Conference on Evolutionary Computation, pp. 786–791 (1996)
17. Mahfoud, S.W.: Niching methods for genetic algorithms. Ph.D. Dissertation, Illinois Genetic Algorithm Laboratory. University of Illinois, Urbana, IL (1995)
18. Mengshoel, O.J., Goldberg, D.E. Probability crowding: deterministic crowding with probabilistic replacement. In: W. Banzhaf (ed.) Proceedings of the International Conference GECCO-1999, pp. 409–416, Orlando, FL (1999)
19. Yin, X., Germay, N. A.: Fast genetic algorithm with sharing scheme using cluster analysis methods in multimodal function optimization. In: Proceedings of the 1993 International Conference on Artificial Neural Networks and Genetic Algorithms, pp. 450–457 (1993)
20. Petrowski, A. A.: Clearing procedure as a niching method for genetic algorithms. In: Proceedings of the 1996 IEEE International Conference on Evolutionary Computation, pp. 798–803, IEEE Press, New York, NY, Nagoya, Japan (1996)
21. Li, J.P., Balazs, M.E., Parks, G.T., Glarkson, P.J.: A species conserving genetic algorithms for multimodal function optimization. Evol. Comput. **10**(3), 207–234 (2002)
22. Lianga, Y., Kwong-Sak, L.: Genetic Algorithm with adaptive elitist-population strategies for multimodal function optimization. Appl. Soft Comput. **11**, 2017–2034 (2011)
23. Wei, L.Y., Zhao, M.: A niche hybrid genetic algorithm for global optimization of continuous multimodal functions. Appl. Math. Comput. **160**(3), 649–661 (2005)
24. Castro, L.N., Zuben, F.J.: Learning and optimization using the clonal selection principle. IEEE Trans. Evol. Comput. **6**, 239–251 (2002)
25. Castro, L.N., Timmis, J.: An artificial immune network for multimodal function optimization. In: Proceedings of the 2002 IEEE International Conference on Evolutionary Computation, pp. 699–704, IEEE Press, New York, NY, Honolulu, Hawaii (2002)
26. Xu, Q., Lei, W., Si, J.: Predication based immune network for multimodal function optimization. Eng. Appl. Artif. Intell. **23**, 495–504 (2010)
27. Kennedy, J., Eberhart, R.C.: Particle swarm optimization. In: Proceedings of the 1995 IEEE International Conference on Neural Networks, vol. 4, pp. 1942–1948 (1995)
28. Liang, Jj, Qin, A.K., Suganthan, P.N.: Comprehensive learning particle swarm optimizer for global optimization of multimodal functions. IEEE Trans. Evol. Comput. **10**(3), 281–295 (2006)
29. Chen, D.B., Zhao, C.X.: Particle swarm optimization with adaptive population size and its application. Appl. Soft. Comput. **9**(1), 39–48 (2009)
30. Sumper, D.: The principles of collective animal behaviour. Philos. Trans. Roy. Soc. Lond. B Biol. Sci. **36**(1465), 5–22 (2006)
31. Petit, O., Bon, R.: Decision-making processes: the case of collective movements. Behav. Process. **84**, 635–647 (2010)

32. Kolpas, A., Moehlis, J., Frewen, T., Kevrekidis, I.: Coarse analysis of collective motion with different communication mechanisms. Math. Biosci. **214**, 49–57 (2008)
33. Couzin, I.: Collective cognition in animal groups. Trends in Cogn. Sci. **13**(1), 36–43 (2008)
34. Couzin, I.D., Krause, J.: Self-organization and collective behavior in vertebrates. Adv. Stud. Behav. **32**, 1–75 (2003)
35. Bode, N., Franks, D., Wood, A.: Making noise: Emergent stochasticity in collective motion. J. Theor. Biol. **267**, 292–299 (2010)
36. Couzi, I., Krause, I., James, R., Ruxton, G., Franks, N.: Collective Memory and Spatial Sorting in Animal Groups. J. Theor. Biol. **218**, 1–11 (2002)
37. Couzin, I.D.: Collective minds. Nature **445**, 715–728 (2007)
38. Bazazi, S., Buhl, J., Hale, J.J., Anstey, M.L., Sword, G.A., Simpson, S.J., Couzin, I.D.: Collective motion and cannibalism in locust migratory bands. Curr. Biol. **18**, 735–739 (2008)
39. Bode, N., Wood, A., Franks, D.: The impact of social networks on animal collective motion. Anim. Behav. **82**(1), 29–38 (2011)
40. Lemasson, B., Anderson, J., Goodwin, R.: Collective motion in animal groups from a neurobiological perspective: the adaptive benefits of dynamic sensory loads and selective attention. J. Theor. Biol. **261**(4), 501–510 (2009)
41. Bourjade, M., Thierry, B., Maumy, M., Petit, O.: Decision-making processes in the collective movements of Przewalski horses families Equus ferus Przewalskii: influences of the environment. Ethology **115**, 321–330 (2009)
42. Banga, A., Deshpande, S., Sumanab, A., Gadagkar, R.: Choosing an appropriate index to construct dominance hierarchies in animal societies: a comparison of three indices. Anim. Behav. **79**(3), 631–636 (2010)
43. Hsu, Y., Earley, R., Wolf, L.: Modulation of aggressive behaviour by fighting experience: mechanisms and contest outcomes. Biol. Rev. **81**(1), 33–74 (2006)
44. Broom, M., Koenig, A., Borries, C.: Variation in dominance hierarchies among group-living animals: modeling stability and the likelihood of coalitions. Behav. Ecol. **20**, 844–855 (2009)
45. Bayly, K.L., Evans, C.S., Taylor, A.: Measuring social structure: a comparison of eight dominance indices. Behav. Process. **73**, 1–12 (2006)
46. Conradt, L., Roper, T.J.: Consensus decision-making in animals. Trends Ecol. Evol. **20**, 449–456 (2005)
47. Okubo, A.: Dynamical aspects of animal grouping. Adv. Biophys. **22**, 1–94 (1986)
48. Reynolds, C.W.: Flocks, herds and schools: a distributed behavioural model. Comp. Graph. **21**, 25–33 (1987)
49. Gueron, S., Levin, S.A., Rubenstein, D.I.: The dynamics of mammalian herds: from individual to aggregations. J. Theor. Biol. **182**, 85–98 (1996)
50. Czirok, A., Vicsek, T.: Collective behavior of interacting self-propelled particles. Phys. A **281**, 17–29 (2000)
51. Ballerini, M.: Interaction ruling collective animal behavior depends on topological rather than metric distance: evidence from a field study. Proc. Natl. Acad. Sci. USA **105**, 1232–1237 (2008)

Chapter 9
Social-Spider Algorithm for Constrained Optimization

9.1 Introduction

The collective intelligent behavior of insect or animal groups in nature such as flocks of birds, colonies of ants, schools of fish, swarms of bees and termites have attracted the attention of researchers. The aggregative conduct of insects or animals is known as swarm behavior. Entomologists have studied this collective phenomenon to model biological swarms while engineers have applied these models as a framework for solving complex real-world problems. This branch of artificial intelligence which deals with the collective behavior of swarms through complex interaction of individuals with no supervision is frequently addressed as swarm intelligence. Bonabeau defined swarm intelligence as "any attempt to design algorithms or distributed problem solving devices inspired by the collective behavior of the social insect colonies and other animal societies" [1]. Swarm intelligence has some advantages such as scalability, fault tolerance, adaptation, speed, modularity, autonomy and parallelism [2].

The key components of swarm intelligence are self-organization and labor division. In a self-organizing system, each of the covered units responds to local stimuli individually and may act together to accomplish a global task, via a labor separation which avoids a centralized supervision. The entire system can thus efficiently adapt to internal and external changes.

Several swarm algorithms have been developed by a combination of deterministic rules and randomness, mimicking the behavior of insect or animal groups in nature. Such methods include the social behavior of bird flocking and fish schooling such as the Particle Swarm Optimization (PSO) algorithm [3], the cooperative behavior of bee colonies such as the Artificial Bee Colony (ABC) technique [4], the social foraging behavior of bacteria such as the Bacterial Foraging Optimization Algorithm (BFOA) [5], the simulation of the herding behavior of krill individuals such as the Krill Herd (KH) method [6], the mating behavior of firefly insects such as the Firefly (FF) method [7] and the emulation of

E. Cuevas et al., *Advances of Evolutionary Computation: Methods and Operators*, Studies in Computational Intelligence 629,
DOI 10.1007/978-3-319-28503-0_9

the lifestyle of cuckoo birds such as the Cuckoo Optimization Algorithm (COA) [8].

In particular, insect colonies and animal groups provide a rich set of metaphors for designing swarm optimization algorithms. Such cooperative entities are complex systems that are composed by individuals with different cooperative-tasks where each member tends to reproduce specialized behaviors depending on its gender [9]. However, most of swarm algorithms model individuals as unisex entities that perform virtually the same behavior. Under such circumstances, algorithms waste the possibility of adding new and selective operators as a result of considering individuals with different characteristics such as sex, task-responsibility, etc. These operators could incorporate computational mechanisms to improve several important algorithm characteristics including population diversity and searching capacities.

Although PSO and ABC are the most popular swarm algorithms for solving complex optimization problems, they present serious flaws such as premature convergence and difficulty to overcome local minima [10, 11]. The cause for such problems is associated to the operators that modify individual positions. In such algorithms, during their evolution, the position of each agent for the next iteration is updated yielding an attraction towards the position of the best particle seen so-far (in case of PSO) or towards other randomly chosen individuals (in case of ABC). As the algorithm evolves, those behaviors cause that the entire population concentrates around the best particle or diverges without control. It does favors the premature convergence or damage the exploration-exploitation balance [12, 13].

The interesting and exotic collective behavior of social insects have fascinated and attracted researchers for many years. The collaborative swarming behavior observed in these groups provides survival advantages, where insect aggregations of relatively simple and "unintelligent" individuals can accomplish very complex tasks using only limited local information and simple rules of behavior [14]. Social-spiders are a representative example of social insects [15]. A social-spider is a spider species whose members maintain a set of complex cooperative behaviors [16]. Whereas most spiders are solitary and even aggressive toward other members of their own species, social-spiders show a tendency to live in groups, forming long-lasting aggregations often referred to as colonies [17]. In a social-spider colony, each member, depending on its gender, executes a variety of tasks such as predation, mating, web design, and social interaction [17, 18]. The web it is an important part of the colony because it is not only used as a common environment for all members, but also as a communication channel among them [19] Therefore, important information (such as trapped prays or mating possibilities) is transmitted by small vibrations through the web. Such information, considered as a local knowledge, is employed by each member to conduct its own cooperative behavior, influencing simultaneously the social regulation of the colony [20].

On the other hand, in real-world applications, most optimization problems are subject to different types of constraints. These kinds of problems are known as constrained optimization problems. A constrained optimization problem is defined as finding parameter vector **x** that minimizes an objective function $J(\mathbf{x})$ subject to inequality and/or equality constraints:

$$\begin{array}{lll} \text{minimize}\, J(\mathbf{x}) & \mathbf{x} = (x_1, \ldots, x_n) \in \Re^n & \\ & l_i \le x_i \le u_i, & i = 1, \ldots, n \\ \text{subject to:} & g_j(\mathbf{x}) \le 0 \quad \text{for} & j = 1, \ldots, q \\ & h_j(\mathbf{x}) = 0 \quad \text{for} & j = q+1, \ldots, m \end{array} \tag{9.1}$$

The objective function J is defined on a search space, S, which is defined as a n-dimensional rectangle in R ($\mathbf{S} \subseteq \mathbf{R}$). Domains of variables are defined by their lower and upper bounds (l_i and u_i). A feasible region $\mathbf{F} \subseteq \mathbf{S}$ is defined by a set of m additional constraints ($m \geq 0$) and **x** is defined on feasible space ($\mathbf{x} \in \mathbf{F} \in \mathbf{S}$). At any point $\mathbf{x} \in \mathbf{F}$, constraints g_j that satisfy $g_j(\mathbf{x}) = 0$ are called active constraints at **x**. By extension, equality constraints h_j are also called active at all points of S [21]. Constrained optimization problems are hard to optimization algorithms and also no single parameter (number of linear, nonlinear and active constraints, the ratio $\rho = |\mathbf{F}|/|\mathbf{S}|$, type of the function, number of variables) is proved to be significant as a major measure of difficulty of the problem [22].

Since most of the optimization algorithms have been primarily designed to address unconstrained optimization problems, constraint handling techniques are usually incorporated in the algorithms in order to direct the search towards the feasible regions of the search space. Methods dealing with the constraints were grouped into four categories [23]: (i) methods based on preserving feasibility of solutions by transforming infeasible solutions to feasible ones with some operators; (ii) methods based on penalty functions which introduce a penalty term into the original objective function to penalize constraint violations in order to solve a constrained problem as an unconstrained one; (iii) methods that make a clear distinction between feasible and infeasible solutions; (iv) other hybrid methods combining evolutionary computation techniques with deterministic procedures for numerical optimization. Considering such mechanisms, several swarm algorithms have been modified to solve constrained optimization problems. Such methods include modified versions of PSO [24], ABC [25] and FF [26]. Each one of the four approaches employed to deal constrains presents advantages such as implementation-easiness and fast calculation whereas adversely posses disadvantages such as tuning difficulties and increase of function evaluations [27]. Therefore, it is reasonable to incorporate a combination of these approaches into a single algorithm in order to increase their potential and to eliminate their drawbacks.

In this chapter, a novel swarm algorithm, called the Social Spider Optimization (SSO-C) is presented for solving constrained optimization tasks. The SSO-C algorithm is based on the simulation of the cooperative behavior of social-spiders. In the presented algorithm, individuals emulate a group of spiders which interact to each other based on the biological laws of the cooperative colony. The algorithm considers two different search agents (spiders): males and females. Depending on gender, each individual is conducted by a set of different evolutionary operators which mimic different cooperative behaviors that are typical in a colony. For constraint handling, the presented algorithm incorporates the combination of two different paradigms in order to direct the search towards feasible regions of the search space. In particular, it has been added: (1) a penalty function which introduces a tendency term into the original objective function to penalize constraint violations in order to solve a constrained problem as an unconstrained one; (2) a feasibility criterion to bias the generation of new individuals toward feasible regions increasing also their probability of getting better solutions. Different to most of existent swarm algorithms, in the presented approach, each individual is modeled considering two genders. Such fact allows not only to emulate in a better realistic way the cooperative behavior of the colony, but also to incorporate computational mechanisms to avoid critical flaws commonly present in the popular PSO and ABC algorithms, such as the premature convergence and the incorrect exploration-exploitation balance. In order to illustrate the proficiency and robustness of the presented approach, it is compared to other similar methods. The comparison examines several standard constrained benchmark functions which are commonly considered in the literature. The results show a high performance of the presented method for searching a global optimum in several benchmark functions and real-world engineering problems.

This chapter is organized as follows. In Sect. 9.2, we introduce basic biological aspects of the algorithm. In Sect. 9.3, the novel SSO-C algorithm and its characteristics are both described. Section 9.4 presents the experimental results and the comparative study. Finally, in Sect. 9.5, conclusions are drawn.

9.2 Biological Fundamentals

Social insect societies are complex cooperative systems that self-organize within a set of constraints. Cooperative groups are better at manipulating and exploiting their environment, defending resources and brood, and allowing task specialization among group members [28, 29]. A social insect colony functions as an integrated unit that not only possesses the ability to operate at a distributed manner, but also to undertake enormous construction of global projects [30]. It is important to acknowledge that global order in social insects can arise as a result of internal interactions among members.

A few species of spiders have been documented exhibiting a degree of social behavior [15]. The behavior of spiders can be generalized into two basic forms: solitary spiders and social spiders [17]. This classification is made based on the level of cooperative behavior that they exhibit [18]. In one side, solitary spiders create and maintain their own web while live in scarce contact to other individuals of the same species. In contrast, social spiders form colonies that remain together over a communal web with close spatial relationship to other group members [19]. Figure 9.1 presents two pictures that show different environments formed by the social spider.

Fig. 9.1 Two pictures a–b that show different environments formed by the social spider

A social spider colony is composed of two fundamental components: its members and the communal web. Members are divided into two different categories: males and females. An interesting characteristic of social-spiders is the highly female-biased population. Some studies suggest that the number of male spiders barely reaches the 30 % of the total colony members [17, 31]. In the colony, each member, depending on its gender, cooperate in different activities such as building and maintaining the communal web, prey capturing, mating and social contact [20]. Interactions among members are either direct or indirect [32]. Direct interactions imply body contact or the exchange of fluids such as mating. For indirect interactions, the communal web is used as a "medium of communication" which conveys important information that is available to each colony member [19]. This information encoded as small vibrations is a critical aspect for the collective coordination among members [20]. Vibrations are employed by the colony members to decode several messages such as the size of the trapped preys, characteristics of the neighboring members, etc. The intensity of such vibrations depend on the weight and distance of the spiders that have produced them.

In spite of the complexity, all the cooperative global patterns in the colony level are generated as a result of internal interactions among colony members [33]. Such internal interactions involve a set of simple behavioral rules followed by each spider in the colony. Behavioral rules are divided into two different classes: social interaction (cooperative behavior) and mating [34].

As a social insect, spiders perform cooperative interaction with other colony members. The way in which this behavior takes place depends on the spider gender. Female spiders which show a major tendency to socialize present an attraction or dislike over others, irrespectively of gender [17]. For a particular female spider, such attraction or dislike is commonly developed over other spiders according to their vibrations which are emitted over the communal web and represent strong colony members [20]. Since the vibrations depend on the weight and distance of the members which provoke them, stronger vibrations are produced either by big spiders or neighboring members [19]. The bigger a spider is, the better it is considered as a colony member. The final decision of attraction or dislike over a determined member is taken according to an internal state which is influenced by several factors such as reproduction cycle, curiosity and other random phenomena [20].

Different to female spiders, the behavior of male members is reproductive-oriented [35]. Male spiders recognize themselves as a subgroup of alpha males which dominate the colony resources. Therefore, the male population is divided into two classes: dominant and non-dominant male spiders [35]. Dominant male spiders have better fitness characteristics (normally size) in comparison to non-dominant. In a typical behavior, dominant males are attracted to the closest female spider in the communal web. In contrast, non-dominant male spiders tend to concentrate upon the center of the male population as a strategy to take advantage of the resources wasted by dominant males [36].

Mating is an important operation that no only assures the colony survival, but also allows the information exchange among members. Mating in a social-spider colony is performed by dominant males and female members [37]. Under such circumstances, when a dominant male spider locates one or more female members within a specific range, it mates with all the females in order to produce offspring [38].

9.3 The Social Spider Optimization (SSO-C) Algorithm for Constrained Optimization

In this chapter, the operational principles from the social-spider colony have been used as guidelines for developing a new swarm optimization algorithm. The SSO-C assumes that entire search space is a communal web, where all the social-spiders interact to each other. In the presented approach, each solution within the search space represents a spider position in the communal web. Every spider receives a weight according to the fitness value of the solution that is symbolized by the social-spider. The algorithm models two different search agents (spiders): males and females. Depending on gender, each individual is conducted by a set of different evolutionary operators which mimic different cooperative behaviors that are commonly assumed within the colony.

An interesting characteristic of social-spiders is the highly female-biased populations. In order to emulate this fact, the algorithm starts by defining the number of female and male spiders that will be characterized as individuals in the search space. The number of females N_f is randomly selected within the range of 65–90 % of the entire population N. Therefore, N_f is calculated by the following equation:

$$\mathrm{N}_f = \mathrm{floor}[(0.9 - \mathrm{rand} \cdot 0.25) \cdot N] \tag{9.2}$$

where rand is a random number between [0,1] whereas floor(·) maps a real number to an integer number. The number of male spiders N_m is computed as the complement between N and N_f . It is calculated as follows:

$$\mathrm{N}_m = N - N_f \tag{9.3}$$

Therefore, the complete population S, composed by N elements, is divided in two sub-groups F and M. The Group F assembles the set of female individuals ($\mathbf{F} = \{\mathbf{f}_1, \mathbf{f}_2, \ldots, \mathbf{f}_{N_f}\}$) whereas M groups the male members ($\mathbf{M} = \{\mathbf{m}_1, \mathbf{m}_2, \ldots, \mathbf{m}_{N_m}\}$), where $\mathbf{S} = \mathbf{F} \cup \mathbf{M}$ ($\mathbf{S} = \{\mathbf{s}_1, \mathbf{s}_2, \ldots, \mathbf{s}_N\}$), such that $\mathbf{S} = \{\mathbf{s}_1 = \mathbf{f}_1, \mathbf{s}_2 = \mathbf{f}_2, \ldots, \mathbf{s}_{N_f} = \mathbf{f}_{N_f}, \mathbf{s}_{N_f+1} = \mathbf{m}_1, \mathbf{s}_{N_f+2} = \mathbf{m}_2, \ldots, \mathbf{s}_N = \mathbf{m}_{N_m}\}$.

9.3.1 Penalty Function and Substituted Function

Since the presented approach aims to solve constrained optimization problems, the original objective function $J(\mathbf{x})$ is replaced by other substituted function $C(\mathbf{x})$ which considers the original objective function $J(\mathbf{x})$ minus a penalty function $P(\mathbf{x})$ that introduces a tendency term to penalize constraint violations produced by x. Therefore, considering the constrained optimization problem defined in Eq. (9.1), the substituted function is defined as follows:

$$\begin{aligned} P(\mathbf{x}) &= \mu \cdot \sum_{i=1}^{q} g_i^2(\mathbf{x}) + v \cdot \sum_{j=q+1}^{m} h_j^2(\mathbf{x}) \\ C(\mathbf{x}) &= J(\mathbf{x}) + P(\mathbf{x}) \end{aligned} \tag{9.4}$$

where μ and v represents the penalty coefficients which weight the relative importance of each kind of constraint. In this work, μ and v are set to 1×10^3 and 10, respectively.

9.3.2 Fitness Assignation

In the biological metaphor, the spider size is the characteristic that evaluates the individual capacity to perform better over its assigned tasks. In the presented approach, every individual (spider) receives a weight w_i which represents the solution quality that corresponds to the spider i (irrespective of gender) of the population S. In order to calculate the weight of every spider the next equation is used:

$$w_i = \frac{worst_{\mathbf{S}} - C(\mathbf{s})}{worst_{\mathbf{S}} - best_{\mathbf{S}}} \tag{9.5}$$

where $C(\mathbf{s}_i)$ is the fitness value obtained by the evaluation of the spider position $\mathbf{s}_i$ with regard to the substituted objective function $C(\cdot)$. The values $worst_{\mathbf{S}}$ and $best_{\mathbf{S}}$ are defined as follows (considering a minimization problem):

$$best_{\mathbf{S}} = \min_{k \in \{1,2,\ldots,N\}} (C(\mathbf{s}_k)) \quad \text{and} \quad worst_{\mathbf{S}} = \max_{k \in \{1,2,\ldots,N\}} (C(\mathbf{s}_k)) \tag{9.6}$$

9.3.3 *Modeling of the Vibrations Through the Communal Web*

The communal web is used as a mechanism to transmit information among the colony members. This information is encoded as small vibrations that are critical for the collective coordination of all individuals in the population. The vibrations depend on the weight and distance of the spider which has generated them. Since the distance is relative to the individual that provokes the vibrations and the member who detects them, members located near to the individual that provokes the vibrations, perceive stronger vibrations in comparison with members located in distant positions. In order to reproduce this process, the vibrations perceived by the individual *i* as a result of the information transmitted by the member *j* are modeled according to the following equation:

$$Vib_{i,j} = w_j \cdot e^{-d_{i,j}^2} \tag{9.7}$$

where the $d_{i,j}$ is the Euclidian distance between the spiders *i* and *j*, such that $d_{i,j} = \|\mathbf{s}_i - \mathbf{s}_j\|$.

Although it is virtually possible to compute perceived-vibrations by considering any pair of individuals, three special relationships are considered within the SSO-C approach:

1. Vibrations $Vibc_i$ are perceived by the individual i ($\mathbf{s}_i$) as a result of the information transmitted by the member c ($\mathbf{s}_c$) who is an individual that has two important characteristics: it is the nearest member to i and possesses a higher weight in comparison to i ($w_c > w_i$).

$$Vibc_i = w_c \cdot e^{-d_{i,c}^2} \tag{9.8}$$

2. The vibrations $Vibb_i$ perceived by the individual i as a result of the information transmitted by the member b ($\mathbf{s}_b$), with b being the individual holding the best weight (best fitness value) of the entire population S, such that $w_b = \max_{k \in \{1,2,\ldots,N\}}(w_k)$.

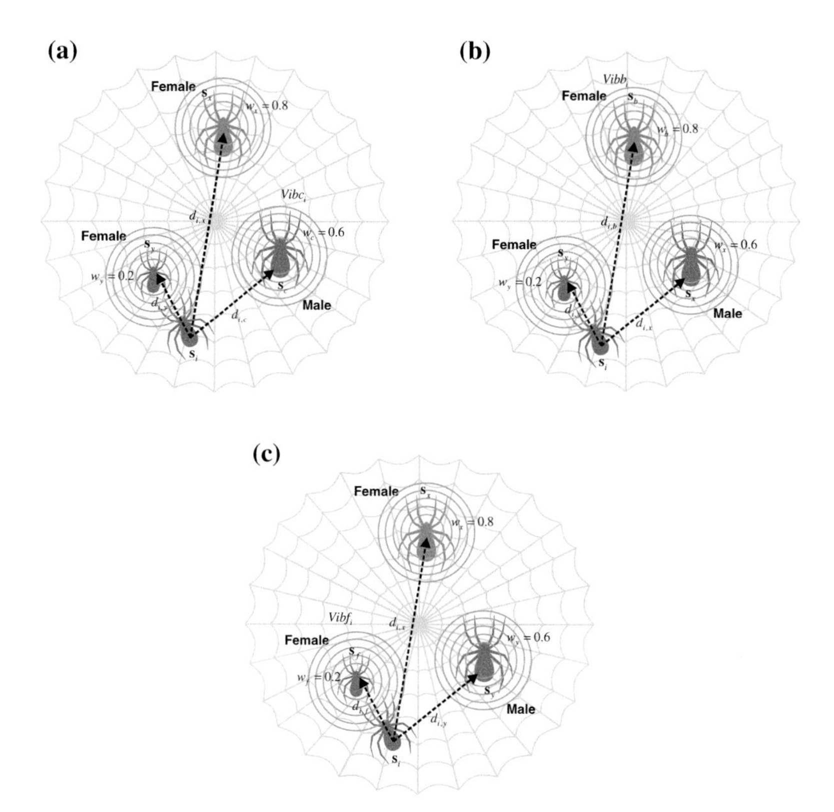

Fig. 9.2 Configuration of each special relation: **a** $Vibc_i$, **b** $Vibb_i$ and **c** $Vibf_i$

$$Vibb_i = w_b \cdot e^{-d_{i,b}^2} \tag{9.9}$$

3. The vibrations $Vibf_i$ perceived by the individual i ($\mathbf{s}_i$) as a result of the information transmitted by the member f ($\mathbf{s}_f$), with f being the nearest female individual to i.

$$Vibf_i = w_f \cdot e^{-d_{i,f}^2} \tag{9.10}$$

Figure 9.2 shows the configuration of each special relationship: (a) $Vibc_i$, (b) $Vibb_i$ and (c) $Vibf_i$.

9.3.4 Initializing the Population

Like other evolutionary algorithms, the SSO-C is an iterative process whose first step is to randomly initialize the entire population (female and male). The algorithm begins by initializing the set S of N spider positions. Each spider position, $\mathbf{f}_i$ or $\mathbf{m}_i$, is a n-dimensional vector containing the parameter values to be optimized. Such values are randomly and uniformly distributed between the pre-specified lower initial parameter bound p_j^{low} and the upper initial parameter bound p_j^{high}, just as it described by the following expressions:

$$\begin{array}{ll} f_{i,j}^0 = p_j^{low} + \text{rand}(0,1) \cdot (p_j^{high} - p_j^{low}) & m_{k,j}^0 = p_j^{low} + \text{rand} \cdot (p_j^{high} - p_j^{low}) \\ i = 1,2,\ldots,N_f; \quad j = 1,2,\ldots,n & k = 1,2,\ldots,N_m; \quad j = 1,2,\ldots,n \end{array} \qquad (9.11)$$

where j, i and k are the parameter and individual indexes respectively whereas zero signals the initial population. The function rand(0,1) generates a random number between 0 and 1. Hence, $f_{i,j}$ is the jth parameter of the ith female spider position.

9.3.5 *Cooperative Operators*

9.3.5.1 Female Cooperative Operator

Social-spiders perform cooperative interaction over other colony members. The way in which this behavior takes place depends on the spider gender. Female spiders present an attraction or dislike over others irrespective of gender. For a particular female spider, such attraction or dislike is commonly developed over other spiders according to their vibrations which are emitted over the communal web. Since vibrations depend on the weight and distance of the members which have originated them, strong vibrations are produced either by big spiders or other neighboring members lying nearby the individual which is perceiving them. The final decision of attraction or dislike over a determined member is taken considering an internal state which is influenced by several factors such as reproduction cycle, curiosity and other random phenomena.

In order to emulate the cooperative behavior of the female spider, a new operator is defined. The operator considers the position change of the female spider i at each iteration. Such position change, which can be of attraction or repulsion, is computed as a combination of three different elements. The first one involves the change in regard to the nearest member to i that holds a higher weight and produces the vibration $Vibc_i$. The second one considers the change regarding the best individual of the entire population S who produces the vibration $Vibb_i$. Finally, the third one incorporates a random movement.

Since the final movement of attraction or repulsion depends on several random phenomena, the selection is modeled as a stochastic decision. For this operation, a uniform random number r_m is generated within the range [0,1]. If r_m is smaller than a threshold PF, an attraction movement is generated; otherwise, a repulsion movement is produced. Therefore, such operator can be modeled as follows:

$$\mathbf{f}_i^{k+1} = \begin{cases} \mathbf{f}_i^k + \alpha \cdot Vibc_i \cdot (\mathbf{s}_c - \mathbf{f}_i^k) + \beta \cdot Vibb_i \cdot (\mathbf{s}_b - \mathbf{f}_i^k) + \delta \cdot (\text{rand} - \frac{1}{2}) \\ \qquad \text{with probability } PF \\ \mathbf{f}_i^k - \alpha \cdot Vibc_i \cdot (\mathbf{s}_c - \mathbf{f}_i^k) - \beta \cdot Vibb_i \cdot (\mathbf{s}_b - \mathbf{f}_i^k) + \delta \cdot (\text{rand} - \frac{1}{2}) \\ \qquad \text{with probability } 1 - PF \end{cases} \tag{9.12}$$

where α, β, δ and rand are random numbers between [0,1] whereas k represents the iteration number. The individual $\mathbf{s}_c$ and $\mathbf{s}_b$ represent the nearest member to i that holds a higher weight and the best individual of the entire population S, respectively.

Under this operation, each particle presents a movement which combines the past position that holds the attraction or repulsion vector over the local best element $\mathbf{s}_c$ and the global best individual $\mathbf{s}_b$ seen so-far. This particular type of interaction avoids the quick concentration of particles at only one point and encourages each particle to search around the local candidate region within its neighborhood ($\mathbf{s}_c$), rather than interacting to a particle ($\mathbf{s}_b$) in a distant region of the domain. The use of this scheme has two advantages. First, it prevents the particles from moving towards the global best position, making the algorithm less susceptible to premature convergence. Second, it encourages particles to explore their own neighborhood thoroughly before converging towards the global best position. Therefore, it provides the algorithm with global search ability and enhances the exploitative behavior of the presented approach.

9.3.5.2 Male Cooperative Operator

According to the biological behavior of the social-spider, male population is divided into two classes: dominant and non-dominant male spiders. Dominant male spiders have better fitness characteristics (usually regarding the size) in comparison to non-dominant. Dominant males are attracted to the closest female spider in the communal web. In contrast, non-dominant male spiders tend to concentrate in the center of the male population as a strategy to take advantage of resources that are wasted by dominant males.

For emulating such cooperative behavior, the male members are divided into two different groups (dominant members D and non-dominant members ND) according to their position with regard to the median member. Male members, with a weight value above the median value within the male population, are considered the dominant individuals D. On the other hand, those under the median value are labeled as non-dominant ND males. In order to implement such computation, the male population M ($\mathbf{M} = \{\mathbf{m}_1, \mathbf{m}_2, \ldots, \mathbf{m}_{N_m}\}$) is arranged according to their weight value in decreasing order. Thus, the individual whose weight w_{N_f+m} is located in the middle is considered the median male member. Since indexes of the male population M in regard to the entire population S are increased by the number of female members N_f, the median weight is indexed by $N_f + m$. According to this, change of positions for the male spider can be modeled as follows:

$$\mathbf{m}_i^{k+1} = \begin{cases} \mathbf{m}_i^k + \alpha \cdot Vibf_i \cdot (\mathbf{s}_f - \mathbf{m}_i^k) + \delta \cdot (\text{rand} - \frac{1}{2}) & \text{if} \quad w_{N_f+i} > w_{N_f+m} \\ \mathbf{m}_i^k + \alpha \cdot \left(\frac{\sum_{h=1}^{N_m} \mathbf{m}_h^k \cdot w_{N_f+h}}{\sum_{h=1}^{N_m} w_{N_f+h}} - \mathbf{m}_i^k \right) & \text{if} \quad w_{N_f+i} \leq w_{N_f+m} \end{cases} \tag{9.13}$$

where the individual $\mathbf{s}_f$ represents the nearest female individual to the male member i whereas $\left(\sum_{h=1}^{N_m} \mathbf{m}_h^k \cdot w_{N_f+h} / \sum_{h=1}^{N_m} w_{N_f+h}\right)$ correspond to the weighted mean of the male population M.

By using this operator, two different behaviors are produced. First, the set D of particles is attracted to others in order to provoke mating. Such behavior allows incorporating diversity into the population. Second, the set ND of particles is attracted to the weighted mean of the male population M. This fact is used to partially control the search process according to the average performance of a sub-group of the population. Such mechanism acts as a filter which avoids that very good individuals or extremely bad individuals influence the search process.

9.3.5.3 Mating Operator

Mating in a social-spider colony is performed by dominant males and the female members. Under such circumstances, when a dominant male $\mathbf{m}_g$ spider ($g \in \mathbf{D}$) locates a set $\mathbf{E}^g$ of female members within a specific range r (range of mating), it mates, forming a new brood $\mathbf{s}_{new}$ which is generated considering all the elements of the set $\mathbf{T}^g$ that, in turn, has been generated by the union $\mathbf{E}^g \cup \mathbf{m}_g$. It is important to emphasize that if the set $\mathbf{E}^g$ is empty, the mating operation is canceled. The range r is defined as a radius which depends on the size of the search space. Such radius r is computed according to the following model:

$$r = \frac{\sum_{j=1}^{n} (p_j^{high} - p_j^{low})}{2 \cdot n} \tag{9.14}$$

In the mating process, the weight of each involved spider (elements of $\mathbf{T}^g$) defines the probability of influence for each individual into the new brood. The spiders holding a heavier weight are more likely to influence the new product, while elements with lighter weight have a lower probability. The influence probability Ps_i of each member is assigned by the roulette method, which is defined as follows:

$$r = \frac{\sum_{j=1}^{n} (p_j^{high} - p_j^{low})}{2 \cdot n}, \quad i \in \mathbf{T}^g \tag{9.15}$$

Once the new spider $\mathbf{s}_{new}$ is formed, it is compared to the spider $\mathbf{s}_{wo}$ holding the worst weight of the colony, where $w_{wo} = \min_{l \in \{1,2,\ldots,N\}} (w_l)$. In order to direct the search towards feasible regions of the search space, a feasibility criterion is incorporated to bias the generation of new spiders toward feasible regions increasing also their probability of getting better solutions. Such feasibility criterion allows choosing one of two different solutions ($\mathbf{s}_{new}$ or $\mathbf{s}_{wo}$) applying the two following rules:

- Between $\mathbf{s}_{new}$ and $\mathbf{s}_{wo}$, it is chosen the spider whose penalty function $P(\mathrm{s})$ present the lower value. Therefore, it is selected $\mathbf{s}_{new}$, if $(P(\mathbf{s}_{new}) < P(\mathbf{s}_{wo}))$ or $\mathbf{s}_{wo}$ if $(P(\mathbf{s}_{new}) > P(\mathbf{s}_{wo}))$.
- If $(P(\mathbf{s}_{new}) = P(\mathbf{s}_{wo}))$, it is selected the spider with better weight. Thus, it is chosen $\mathbf{s}_{new}$, if $(w_{new} \geq w_{wo})$, otherwise $\mathbf{s}_{wo}$ is considered.

In case of $\mathbf{s}_{new}$ would be selected, the new spider $\mathbf{s}_{new}$ assumes the gender and index from the replaced spider $\mathbf{s}_{wo}$. Such fact assures that the entire population S maintains the original rate between female and male members.

In order to demonstrate the mating operation, Fig. 9.3 illustrates a simple optimization problem which can be stated as follows:

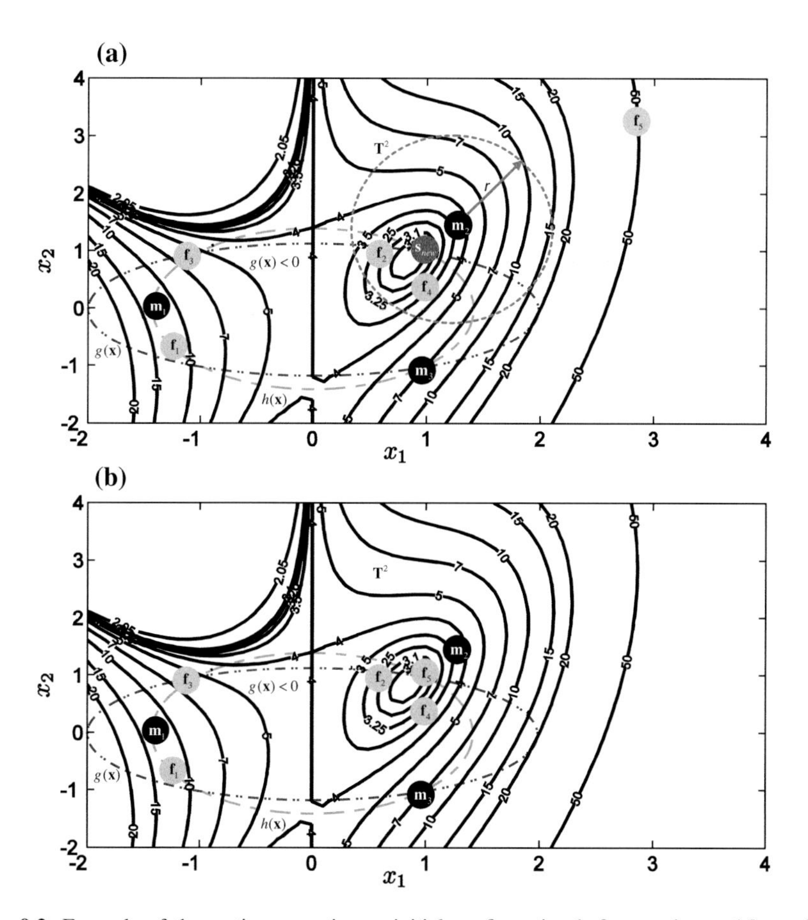

Fig. 9.3 Example of the mating operation: **a** initial configuration before mating and **b** configuration after the mating operation

Table 9.1 Data for constructing the new spider $\mathbf{s}_{new}$ through the roulette method

Spider		Position	w_i	Ps_i	$P(\mathbf{s})$	Roulette
$\mathbf{s}_1$	$\mathbf{f}_1$	(−1.2,−0.8)	0.88	–	0	
$\mathbf{s}_2$	$\mathbf{f}_2$	(0.7,1)	0.83	0.40	18.85	
$\mathbf{s}_3$	$\mathbf{f}_3$	(−1.1,0.9)	0.95	–	8.10	
$\mathbf{s}_4$	$\mathbf{f}_4$	(1,0.3)	0.74	0.36	474	
$\mathbf{s}_5$	$\mathbf{f}_5$	(2.8,3.3)	0	–	86,002	
$\mathbf{s}_6$	$\mathbf{m}_1$	(−1.4,0)	0.79	–	260	
$\mathbf{s}_7$	$\mathbf{m}_2$	(1.1,1.5)	0.49	0.24	1002	
$\mathbf{s}_8$	$\mathbf{m}_3$	(1,−1)	1	–	0	
$\mathbf{s}_{new}$		(1,1)	1.2	–	0	

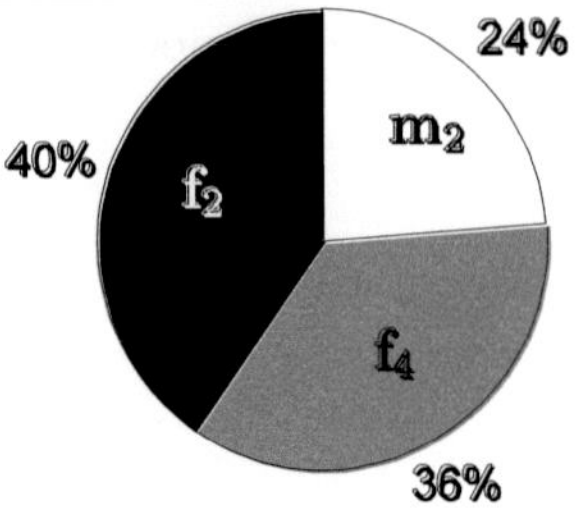

$$\begin{array}{ll} \text{minimize} & J(\mathbf{x}) = x_1^4 - 2x_1^2 x_2 + x_1^2 + x_1 x_2^2 - 2x_1 + 4 \\ \text{subject to:} & -2 \leq x_1 \leq 4 \quad -2 \leq x_2 \leq 4 \\ & g(\mathbf{x}) = 0.25x_1^2 + 0.75x_2^2 - 1 \leq 0 \\ & h(\mathbf{x}) = x_1^2 + x_2^2 - 2 = 0 \end{array} \tag{9.16}$$

As an example, it is assumed a population S of eight different 2-dimensional members ($N = 8$), five females ($N_f = 5$) and three males ($N_m = 3$). Figure 9.3a shows the initial configuration of the presented example with three different female members $\mathbf{f}_2(\mathbf{s}_2)$ and $\mathbf{f}_4(\mathbf{s}_4)$ constituting the set $\mathbf{E}^2$ which is located inside of the influence range r of a dominant male $\mathbf{m}_2(\mathbf{s}_7)$. Then, the new candidate spider $\mathbf{s}_{new}$ is generated from the elements $\mathbf{f}_2$, $\mathbf{f}_4$ and $\mathbf{m}_2$ which constitute the set $\mathbf{T}^2$. Therefore, the value of the first decision variable $s_{new,1}$ for the new spider is chosen by means of the roulette mechanism considering the values already existing from the set $\{f_{2,1}, f_{4,1}, m_{2,1}\}$. The value of the second decision variable $s_{new,2}$ is also chosen in the same manner. Table 9.1 shows the data for constructing the new spider through the roulette method. Once the new spider $\mathbf{s}_{new}$ is formed, the feasibility criterion is applied, considering the worst member $\mathbf{f}_5$ that is present in the population S. Since the penalty function $P(\mathbf{s}_{new})$ is lesser than $P(\mathbf{s}_5)$, $\mathbf{f}_5$ is replaced by $\mathbf{s}_{new}$. Therefore, $\mathbf{s}_{new}$ assumes the same gender and index from $\mathbf{f}_5$. Figure 9.3b shows the configuration of S after the mating process.

Under this operation, new generated particles locally exploit the search space inside the mating range in order to find better individuals.

9.3.6 Computational Procedure

The computational procedure for the presented algorithm can be summarized as follows:

Step 1	Considering N as the total number of n-dimensional colony members, define the number of male N_m and females N_f spiders in the entire population S $N_f = \text{floor}[(0.9 - \text{rand} \cdot 0.25) \cdot N]$ and $N_m = N - N_f$, where rand is a random number between [0,1] whereas floor(·) maps a real number to an integer number

Step 2	Initialize randomly he female ($\mathbf{F} = \{\mathbf{f}_1, \mathbf{f}_2, \ldots, \mathbf{f}_{N_f}\}$) and male ($\mathbf{M} = \{\mathbf{m}_1, \mathbf{m}_2, \ldots, \mathbf{m}_{N_m}\}$) members where $\mathbf{S} = \{\mathbf{s}_1 = \mathbf{f}_1, \mathbf{s}_2 = \mathbf{f}_2, \ldots, \mathbf{s}_{N_f} = \mathbf{f}_{N_f}, \mathbf{s}_{N_f+1} = \mathbf{m}_1, \mathbf{s}_{N_f+2} = \mathbf{m}_2, \ldots, \mathbf{s}_N = \mathbf{m}_{N_m}\}$ and calculate the radius of mating
	$r = \frac{\sum_{j=1}^{n} (p_j^{high} - p_j^{low})}{2 \cdot n}$
	for (i = 1; $i < N_f$ + 1; i++)
	for (j = 1; j < n+1; j++)
	$f_{i,j}^0 = p_j^{low} + \text{rand}(0,1) \cdot (p_j^{high} - p_j^{low})$
	end for
	end for
	for (k = 1; $k < N_m$ + 1; k++)
	for (j = 1; j < n + 1; j++)
	$m_{k,j}^0 = p_j^{low} + \text{rand} \cdot (p_j^{high} - p_j^{low})$
	end for
	end for

Step 3	Calculate the weight of every spider of S (Sect. 9.3.1)
	for (i = 1; $i < N$ + 1; i++)
	$w_i = \frac{worst_{\mathbf{S}} - C(\mathbf{s})}{worst_{\mathbf{S}} - best_{\mathbf{S}}}$
	where $C(\cdot)$ represent the substituted function, $best_{\mathbf{S}} = \min_{k \in \{1,2,\ldots,N\}} (J(\mathbf{s}_k))$ and
	$worst_{\mathbf{S}} = \max_{k \in \{1,2,\ldots,N\}} (J(\mathbf{s}_k))$
	end for

Step 4	Move female spiders according to the female cooperative operator (Sect. 9.3.5)
	for (i = 1; $i < N_f$ + 1; i++)
	Calculate $Vibc_i$ and $Vibb_i$ (Sect. 9.3.3)
	If ($r_m < PF$); where $r_m \in \text{rand}(0,1)$
	$\mathbf{f}_i^{k+1} = \mathbf{f}_i^k + \alpha \cdot Vibc_i \cdot (\mathbf{s}_c - \mathbf{f}_i^k) + \beta \cdot Vibb_i \cdot (\mathbf{s}_b - \mathbf{f}_i^k) + \delta \cdot (\text{rand} - \frac{1}{2})$
	else if
	$\mathbf{f}_i^{k+1} = \mathbf{f}_i^k - \alpha \cdot Vibc_i \cdot (\mathbf{s}_c - \mathbf{f}_i^k) - \beta \cdot Vibb_i \cdot (\mathbf{s}_b - \mathbf{f}_i^k) + \delta \cdot (\text{rand} - \frac{1}{2})$
	end if
	end for

(continued)

(continued)

Step 5	Move the male spiders according to the male cooperative operator (Sect. 9.3.5)
	Find the median male individual (w_{N_f+m}) from **M**
	for (i = 1; i < N_m + 1; i++)
	Calculate $Vibf_i$ (Sect. 3.1.3)
	If ($w_{N_f+i} > w_{N_f+m}$)
	$\mathbf{m}_i^{k+1} = \mathbf{m}_i^k + \alpha \cdot Vibf_i \cdot (\mathbf{s}_f - \mathbf{m}_i^k) + \delta \cdot \left(\text{rand} - \frac{1}{2}\right)$
	Else if
	$\mathbf{m}_i^{k+1} = \mathbf{m}_i^k + \alpha \cdot \left(\frac{\sum_{h=1}^{N_m} \mathbf{m}_h^k \cdot w_{N_f+h}}{\sum_{h=1}^{N_m} w_{N_f+h}} - \mathbf{m}_i^k\right)$
	end if
	end for

Step 6	Perform the mating operation (Sect. 9.3.6)
	for (i = 1; i < N_m + 1; i++)
	If ($\mathbf{m}_i \in \mathbf{D}$)
	Find $\mathbf{E}^i$
	If ($\mathbf{E}^i$ is not empty)
	Form $\mathbf{s}_{new}$ using the roulette method
	Apply the feasibility criterion
	In case of $\mathbf{s}_{new}$ would be selected
	$\mathbf{s}_{new}$ assumes the gender and index from $\mathbf{s}_{wo}$
	end if
	end if
	end for

Step 7	If the stop criteria is met, the process is finished; otherwise, go back to Step 3

9.3.7 *Discussion About the SSO-C Algorithm*

Evolutionary algorithms (EA) have been widely employed for solving complex optimization problems. These methods are found to be more powerful than conventional methods based on formal logics or mathematical programming [39]. In an EA algorithm, search agents have to decide whether to explore unknown search positions or to exploit already tested positions in order to improve their solution quality. Pure exploration degrades the precision of the evolutionary process but increases its capacity to find new potential solutions. On the other hand, pure exploitation allows refining existent solutions but adversely drives the process to local optimal solutions. Therefore, the ability of an EA to find a global optimal solution depends on its capacity to find a good balance between the exploitation of found-so-far elements and the exploration of the search space [40]. So far, the exploration–exploitation dilemma has been an unsolved issue within the framework of evolutionary algorithms.

EA defines individuals with the same property, performing virtually the same behavior. Under these circumstances, algorithms waste the possibility to add new and selective operators as a result of considering individuals with different characteristics. These operators could incorporate computational mechanisms to improve several important algorithm characteristics such as population diversity or searching capacities.

On the other hand, PSO and ABC are the most popular swarm algorithms for solving complex optimization problems. However, they present serious flaws such as premature convergence and difficulty to overcome local minima [10, 11]. Such problems arise from operators that modify individual positions. In such algorithms, the position of each agent in the next iteration is updated yielding an attraction towards the position of the best particle seen so-far (in case of PSO) or any other randomly chosen individual (in case of ABC). Such behaviors produce that the entire population concentrates around the best particle or diverges without control as the algorithm evolves, either favoring the premature convergence or damaging the exploration-exploitation balance [12, 13].

Different to other EA, at SSO-C each individual is modeled considering the gender. Such fact allows incorporating computational mechanisms to avoid critical flaws such as premature convergence and incorrect exploration-exploitation balance commonly present in both, the PSO and the ABC algorithm. From an optimization point of view, the use of the social-spider behavior as a metaphor introduces interesting concepts in EA: the fact of dividing the entire population into different search-agent categories and the employment of specialized operators that are applied selectively to each of them. By using this framework, it is possible to improve the balance between exploitation and exploration, yet preserving the same population, i.e. individuals who have achieved efficient exploration (female spiders) and individuals that verify extensive exploitation (male spiders). Furthermore, the social-spider behavior mechanism introduces an interesting computational scheme with three important particularities: first, individuals are separately processed according to their characteristics. Second, operators share the same communication mechanism allowing the employment of important information of the evolutionary process to modify the influence of each operator. Third, although operators modify the position of only an individual type, they use global information (positions of all individual types) in order to perform such modification. Figure 9.4 presents a schematic representation of the algorithm-data-flow. According to Fig. 9.4, the female cooperative and male cooperative operators process only female or male individuals, respectively. However, the mating operator modifies both individual types.

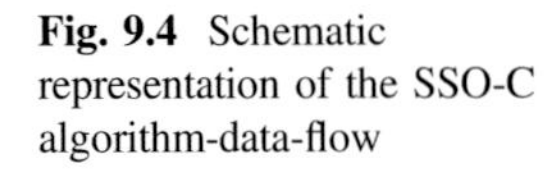

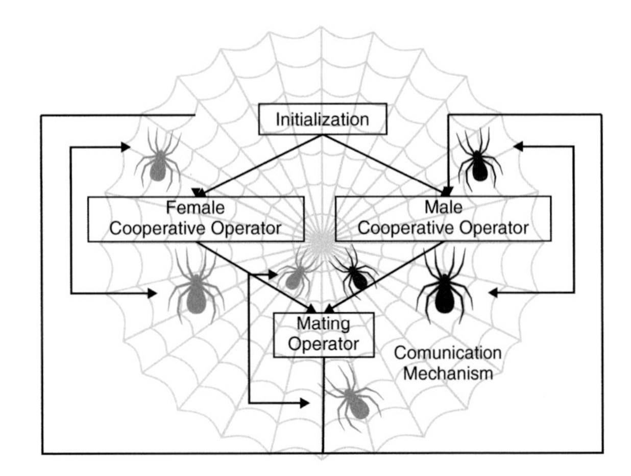

Fig. 9.4 Schematic representation of the SSO-C algorithm-data-flow

9.4 Experimental Results

In order to evaluate how well SSO-C performs on finding the global minimum of the benchmark constrained optimization problems, two categories of problems are considered in this chapter. The first category includes the first 8 well-studied problems (Problems J_1 –J_8) of CEC 2006 test suite [41]. The mathematical expressions of these problems are provided in Table 9.2. The second category includes a set of three well-studied real-world engineering optimization problems.

The SSO-C algorithm have been applied to the benchmark problems whose results have been compared to those produced by modified versions of PSO [24], ABC [25] and FF [26] which have been adapted to work over constrained optimization problems. Such approaches are considered as the most popular swarm algorithms for many optimization applications. The parameter setting for each algorithm in the comparison is described as follows:

1. PSO: The parameters are set to $M = 250$, $G_{\max} = 300$, $c_1 = 2$ and $c_2 = 2$; besides, the weight factor decreases linearly from 0.9 to 0.4 [24].
2. ABC: The algorithm has been implemented using the guidelines provided by its own Ref. [25], using $MR = 0.8$, $sn = 40$ and $MCN = 6000$.
3. FF [26]: The parameters are set to Number of fireflies = 25, $\alpha = 0.001$, $q = 1.5$ and iteration number = 2000.
4. SSO-C: Once it has been determined experimentally, they are kept for all experiments in this section. Such parameters are set to $N = 50$, $PF = 0.7$, iteration number = 500.

Table 9.2 Constrained test functions used in the experimental study

Problem J_5	
Minimize	$J_1(\mathbf{x}) = 5\sum_{d=1}^{4} x_d - 5\sum_{d=1}^{4} x_d^2 - \sum_{d=5}^{13} x_d$
Subject to:	$g_1(\mathbf{x}) = 2x_1 + 2x_2 + x_{10} + x_{11} - 10 \leq 0$
	$g_2(\mathbf{x}) = 2x_1 + 2x_3 + x_{10} + x_{12} - 10 \leq 0$
	$g_3(\mathbf{x}) = 2x_2 + 2x_3 + x_{11} + x_{12} - 10 \leq 0$
	$g_4(\mathbf{x}) = -8x_1 + x_{10} \leq 0$
	$g_5(\mathbf{x}) = -8x_2 + x_{11} \leq 0$
	$g_6(\mathbf{x}) = -8x_3 + x_{12} \leq 0$
	$g_7(\mathbf{x}) = -2x_4 - x_5 + x_{10} \leq 0$
	$g_8(\mathbf{x}) = -2x_6 - x_7 + x_{11} \leq 0$
	$g_9(\mathbf{x}) = -2x_8 - x_9 + x_{12} \leq 0$
With boundary conditions $0 \leq x_d \leq 1$ $(d = 1, \ldots, 9, 13)$,$0 \leq x_d \leq 100(d = 10, 11, 12)$. The optimum objective value is $J_1(\mathbf{x}^*) = -15$	
Problem J_2	
Minimize	$J_2(\mathbf{x}) = -\left\lvert \left(\sum_{d=1}^{n} \cos^4(x_d) - 2\prod_{d=1}^{n} \cos^2(x_d)\right) \Big/ \left(\sqrt{\sum_{d=1}^{n} d \cdot x_d^2}\right) \right\rvert$
Subject to:	$g_1(\mathbf{x}) = 0.75 - \prod_{d=1}^{n} x_d \leq 0$
	$g_2(\mathbf{x}) = \sum_{d=1}^{n} x_d - 0.75n \leq 0$
where n = 20 and $0 \leq x_d \leq 1$. The best objective value is $J_2(\mathbf{x}^*) = -0.803619$	
Problem J_3	
Minimize	$J_3(\mathbf{x}) = 5.3578547x_3^2 + 0.8356891x_1x_5 + 37.293239x_1 - 40792.141$
Subject to:	$g_1(\mathbf{x}) = 85.334407 + 0.0056858x_2x_5 + 0.0006262x_1x_4 - 0.00022053x_3x_5 - 92 \leq 0$
	$g_2(\mathbf{x}) = -85.334407 - 0.0056858x_2x_5 - 0.0006262x_1x_4 + 0.0022053x_3x_5 \leq 0$
	$g_3(\mathbf{x}) = 80.51249 + 0.0071317x_2x_5 + 0.0029955x_1x_2 + 0.0021813x_3^2 - 110 \leq 0$
	$g_4(\mathbf{x}) = -80.51249 - 0.0071317x_2x_5 - 0.0029955x_1x_2 - 0.0021813x_3^2 + 90 \leq 0$
	$g_5(\mathbf{x}) = 9.300961 + 0.0047026x_3x_5 + 0.0012547x_1x_3 + 0.0019085x_3x_4 - 25 \leq 0$
	$g_6(\mathbf{x}) = -9.300961 - 0.0047026x_3x_5 - 0.0012547x_1x_3 - 0.0019085x_3x_4 + 20 \leq 0$
With boundary conditions $78 \leq x_1 \leq 102$, $33 \leq x_2 \leq 45$, and $27 \leq x_d \leq 45$ (d = 3, 4, 5). The optimum objective value is $J_3(\mathbf{x}^*) = -30{,}665.53867$	
Problem J_4	
Minimize	$J_4(\mathbf{x}) = (x_1 - 10)^3 + (x_2 - 20)^3$
Subject to:	$g_1(\mathbf{x}) = -(x_1 - 5)^2 - (x_2 - 5)^2 + 100 \leq 0$
	$g_2(\mathbf{x}) = (x_1 - 6)^2 + (x_2 - 5)^2 - 82.81 \leq 0$
With boundary conditions $13 \leq x_1 \leq 100$ and $0 \leq x_2 \leq 100$. The optimum objective value is $J_4(\mathbf{x}^*) = -6961.81387$	
Problem J_5	
Minimize	$J_5(\mathbf{x}) = x_1^2 + x_2^2 + x_1x_2 - 14x_1 - 16x_2 + (x_3 - 10)^2 + 4(x_4 - 5)^2 + (x_5 - 3)^2 + 2(x_6 - 1)^2 + 5x_7^2$ $\cdots + 7(x_8 - 11)^2 + 2(x_9 - 10)^2 + (x_{10} - 7)^2 + 45$
Subject to:	$g_1(\mathbf{x}) = -105 + 4x_1 + 5x_2 - 3x_7 + 9x_8 \leq 0$
	$g_2(\mathbf{x}) = 10x_1 - 8x_2 - 17x_7 + 2x_8 \leq 0$
	$g_3(\mathbf{x}) = -8x_1 + 2x_2 + 5x_9 - 2x_{10} - 12 \leq 0$
	$g_4(\mathbf{x}) = 3(x_1 - 2)^2 + 4(x_2 - 3)^2 + 2x_3^2 - 7x_4 - 120 \leq 0$
	$g_5(\mathbf{x}) = 5x_1^2 + 8x_2 + (x_3 - 6)^2 - 2x^4 - 40 \leq 0$
	$g_6(\mathbf{x}) = x_1^2 + 2(x_2 - 2)^2 - 2x_1x_2 + 14x_5 - 6x_6 \leq 0$
	$g_7(\mathbf{x}) = 0.5(x_1 - 8)^2 + 2(x_2 - 4)^2 + 3x_5^2 - x_6 - 30 \leq 0$
	$g_8(\mathbf{x}) = -3x_1 + 6x_2 + 12(x_9 - 8)^2 - 7x_{10} \leq 0$

(continued)

Table 9.2 (continued)

Problem J_5	
where $-10 \le x_d \le 10$ (d = 1, …, 10). The global objective value is $J_5(\mathbf{x}^*) = 24.306209$	
Problem J_6	
Minimize	$J_6(\mathbf{x}) = x_1^2 + (x_2 - 1)^2$
Subject to:	$h_1(\mathbf{x}) = x_2 - x_1^2 = 0$
With $-1 \le x_d \le 1$ (d = 1,2). The global objective value is $J_6(\mathbf{x}^*) = 0.7499$	
Problem J_7	
Minimize	$J_7(\mathbf{x}) = e^{x_1 x_2 x_3 x_4 x_5}$
Subject to:	$h_1(\mathbf{x}) = x_1^2 + x_2^2 + x_3^2 + x_4^2 + x_5^2 - 10 = 0$
	$h_2(\mathbf{x}) = x_2 x_3 - 5x_4 x_5 = 0$ $h_3(\mathbf{x}) = x_1^3 + x_2^3 + 1 = 0$
With boundary conditions $-2.3 \le x_d \le 2.3$ (d = 1, 2), $-3.2 \le x_d \le 3.2$ (d = 3, 4, 5), and $10 \le x_d \le 1000$ (d = 4, …, 8). The optimum objective value is $J_7(\mathbf{x}^*) = 0.0539415$	
Problem J_8	
Minimize	$J_8(\mathbf{x}) = 1000 - x_1^2 - 2x_2^2 - x_3^2 - x_1 x_2 - x_1 x_3$
Subject to:	$h_1(\mathbf{x}) = x_1^2 + x_2^2 + x_3^2 - 25 = 0$
	$h_2(\mathbf{x}) = 8x_1 + 14x_2 + 7x_3 - 56 = 0$
where $0 \le x_d \le 10$, (d = 1, 2, 3). The optimum objective value is $J_8(\mathbf{x}^*) = 961.715022289961$	

9.4.1 Comparisons on J_1–J_8 Benchmark Problems

This section evaluates the performance of the presented algorithm on 8 well-defined constrained optimization problems [41]. These problems are widely recognized for having different properties and include various types of objective functions (i.e., linear, nonlinear, and quadratic) with different number of decision variables (n) and linear/nonlinear equalities/inequalities. All equality constraints $h_j(\mathbf{x}) = 0$, are converted into inequality constraints $|h_j(\mathbf{x})| \le \varepsilon$ with $\varepsilon = 0.0001$. The simulations have been conducted in MATLAB and executed 30 times independently on each problem whereas it is reported the statistical features of the results obtained.

Table 9.3 summarizes the results obtained by all algorithms considering the benchmark problems J_1–J_8. In this table, results are based on the best (B), mean (M), worst (W) and standard deviation (SD) of the lowest function values obtained by each algorithm. As can be seen from the results of Table 9.3, SSO-C always reaches the global optimum of all problems whereas in terms of the average and worst performance, SSO-C presents the best stability. Similarly, the presented algorithm obtains the best precision, since the standard deviation (SD) presents the lower values.

Table 9.3 Minimization results of benchmark functions of Table 9.2

Function	Optimal	Index	PSO	ABC	FF	SSO-C
J_1	−15.000	B	−15.000	−15.000	14.999	−15.000
		M	−15.000	−15.000	14.988	−15.000
		W	−15.000	−15.000	14.798	−15.000
		SD	0	0	6.4E−07	0
J_2	−0.803619	B	−0.80297	−0.803388	−0.803601	−0.803619
		M	−0.79010	−0.790148	−0.785238	−0.801563
		W	−0.76043	−0.756986	−0.751322	0.792589
		SD	1.2E−02	1.3E−02	1.67E−03	3.5E−05
J_3	−30,665.539	B	−30,665.501	−30,665.539	−30,664.322	−30,665.539
		M	−30,662.821	−30,664.923	−30,662.032	−30,665.538
		W	−30,650.432	−30,659.131	−30,648.974	−30,665.147
		SD	5.2E−02	8.2E−02	5.2E−02	1.1E−04
J_4	−6961.814	B	−6961.728	−6961.814	−6959.987	−6961.814
		M	−6958.369	−6958.022	−6950.114	−6961.008
		W	−6942.085	−6955.337	−6947.626	−6960.918
		SD	6.7E−02	2.1E−02	3.8E−02	1.1E−03
J_5	24.306	B	24.327	24.48	23.97	24.306
		M	24.475	26.58	28.54	24.306
		W	24.843	28.40	30.14	24.306
		SD	1.32E−01	1.14	2.25	4.95E-05
J_6	−0.7499	B	−0.7499	−0.7499	−0.7497	−0.7499
		M	−0.7490	−0.7495	−0.7491	−0.7499
		W	−0.7486	−0.7490	−0.7479	−0.7499
		SD	1.2E−03	1.67E−03	1.5E−03	4.1E−09
J_7	0.0539415	B	0.05411	0.05394	0.05410	0.05394
		M	0.05416	0.05398	0.05417	0.05394
		W	0.05421	0.05411	0.05425	0.05394
		SD	1.35E−03	2.2E−03	3.1E−03	6.3E−09
J_8	961.715022	B	962.2132	961.9821	963.6281	961.9821
		M	963.9251	962.6421	965.4281	961.9987
		W	965.0251	964.2417	969.3217	962.0078
		SD	3.21	2.4	3.7	1.2

9.4.2 *Comparisons on Mechanical Engineering Design Optimization Problems*

In order to assess the performance of SSO-C on complex real word engineering problems, three well-studied problems are adopted from literature. These problems are: the tension/compression spring design optimization problem, the welded beam design optimization problem and the speed reducer design optimization problem. We compare the performance of with those produced by modified versions of PSO, ABC and FF.

9.4.2.1 Tension/Compression Spring Design Optimization Problem

The aim in this problem is to minimize the weight of a tension/compression spring (see Fig. 9.5) subject to constraints on minimum deflection, shear stress, surge frequency and limits on outside diameter [42]. This problem has three continuous variables defined as: the wire diameter (x_1), the mean coil diameter (x_2) and the number of active coils (x_3). The constrained optimization problem is defined in Table 9.4.

Results are tabulated in Table 9.5. It can be verified that the SSO-C presents a good performance whereas it maintains an acceptable stability. Moreover, in terms of the mean, worst, and standard deviation values, SSO-C dominates the 3 algorithms. However, there are other algorithms that can obtain along with SSO-C similar values, in terms of the best performance. The best solution obtained by SSO-C is $\mathbf{x}^* =(0.051689061657295, 0.356717753621158, 11.288964941289167)$ with $J_{P1}(\mathbf{x}^*) = 0.0126652327883194$.

9.4.2.2 Welded Beam Design Optimization Problem

This problem is aimed to minimize the cost of a beam subject to constraints on shear stress (τ), bending stress in the beam (σ), buckling load on the bar (P_c) and end deflection of the beam (δ). The design variables are the thickness of the weld (x_1), length of the weld (x_2), width of the beam (x_3), and the beam thickness (x_4).

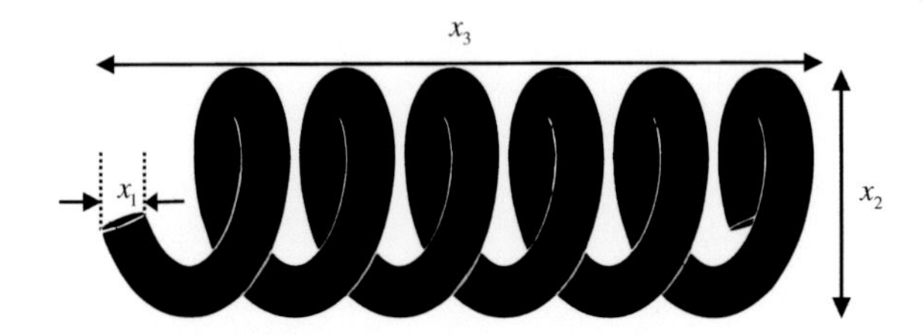

Fig. 9.5 A tension/compression string and its design features

Table 9.4 Tension/compression spring design optimization problem

Problem: Tension/compression spring design optimization problem	
Minimize	$J_{P1}(\mathbf{x}) = (x_3 + 2)x_2x_1^2$
Subject to:	$g_1(\mathbf{x}) = 1 - \frac{x_2^3x_3}{71{,}785x_1^4} \leq 0$
	$g_2(\mathbf{x}) = \frac{4x_2^2 - x_1x_2}{12{,}566x_2x_1^3 - x_1^4} + \frac{1}{5108x_1^2} \leq 0$
	$g_3(\mathbf{x}) = 1 - \frac{140.45x_1}{x_2^2x_3} \leq 0$
	$g_4(\mathbf{x}) = \frac{x_2 + x_1}{1.5} - 1 \leq 0$
With boundary conditions $0.05 \leq x_1 \leq 2, 0.25 \leq x_2 \leq 1.3$ and $2 \leq x_3 \leq 15$	

Table 9.5 Statistical features of the results obtained by various algorithms on the spring design problem

	B	M	W	SD
PSO	0.01285757491586	0.014863120662235	0.019145260276060	0.001261916628168
ABC	0.01266523390012	0.012850718301673	0.013210405561696	0.000118451102725
FF	0.01266523390345	0.012930718743571	0.013420409563257	0.001453639618101
SSO-C	0.01266523278831	0.012764888178309	0.012867916581611	0.000092874978053

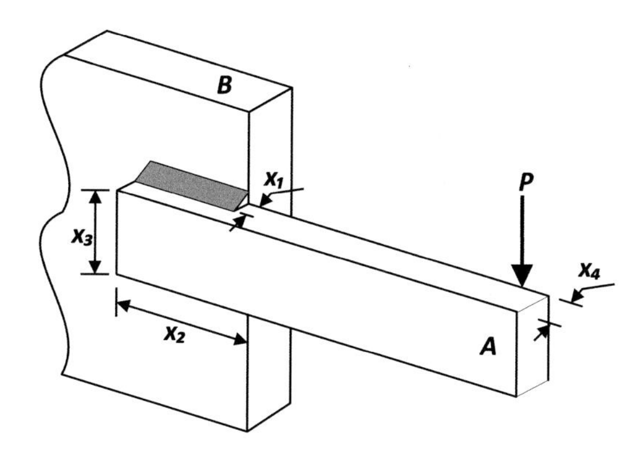

Fig. 9.6 A welded beam and its design features

Figure 9.6 depicts a welded beam together with its design features. The mathematical formulation for this problem [43] is described in Table 9.6.

The experiment compares the SSO-C to other algorithms such as PSO and ABC and FF. The results for 30 runs are reported in Table 9.7. According to this table, SSO-C delivers better results than PSO, ABC in terms of the mean, worst, and standard deviation values. In particular, the test remarks the largest difference in performance which is directly related to a better trade-off between exploration and exploitation. The best solution obtained by SSO-C is $\mathbf{x}^* = $(0.205729639786079, 3.470488665628002, 9.036623910357633, 0.205729639786080) with $J_{P2}(\mathbf{x}^*) =$ 1.72485230 8597365.

Table 9.6 Welded beam design optimization problem

Problem: welded beam design optimization problem	
Minimize	$J_{P2}(\mathbf{x}) = 1.10471x_2x_1^2 + 0.04811x_3x_4(14 + x_2)$
Subject to:	$g_1(\mathbf{x}) = \tau(X) - 13{,}600 \leq 0$
	$g_2(\mathbf{x}) = \sigma(X) - 30{,}000 \leq 0$
	$g_3(\mathbf{x}) = x_1 - x_4 \leq 0$
	$g_4(\mathbf{x}) = 0.10471x_1^2 + 0.04811x_3x_4(14 + x_2) - 5 \leq 0$
	$g_5(\mathbf{x}) = 0.125 - x_1 \leq 0$
	$g_6(\mathbf{x}) = \delta(X) - 0.25 \leq 0$
	$g_7(\mathbf{x}) = 6000 - P_c(X) \leq 0$
where:	$\tau(X) = \sqrt{a^2 + 2ab(x_2/2R) + b^2}$
	$a = \frac{6000}{\sqrt{2}x_1x_2}\ b = \frac{6000\cdot\left(14 + \frac{x_2}{2}\right)\cdot R}{2\sqrt{2}x_1x_2\left[\frac{x_2^2}{12} + \left(\frac{x_1 + x_3}{2}\right)^2\right]}\ R = \sqrt{\frac{x_2^2}{4} + \left(\frac{x_1 + x_3}{2}\right)^2}$
	$\sigma(X) = \frac{504{,}000}{x_4x_3^2}\ \delta(X) = \frac{65{,}856{,}000}{(30 \times 10^6)x_4x_3^3}$
	$P_c(X) = \frac{4.013(30 \times 10^6)}{196}\sqrt{\frac{x_3^2x_4^6}{36}}\left[1 + \left(x_3\frac{\sqrt{\frac{30 \times 10^6}{4(12 \times 10^6)}}}{28}\right)\right]$
With boundary conditions $0.1 \leq x_1 \leq 2, 0.1 \leq x_2 \leq 10, 0.1 \leq x_3 \leq 10$ and $0.1 \leq x_4 \leq 2$	

Table 9.7 Statistical features of the results obtained by various algorithms on the welded beam design optimization problem

	B	M	W	SD
PSO	1.8464084393	2.011146016	2.237388681	0.108512649
ABC	1.7981726165	2.167357771	2.887044413	0.254266058
FF	1.7248541012	2.197401062	2.931001383	0.195264102
SSO-C	1.7248523085	1.746461619	1.799331766	0.025729853

9.4.2.3 Speed Reducer

A speed reducer is part of the gear box of mechanical systems, and it is also used for many other types of applications. The design of a speed reducer is considered as a challenging optimization problem in mechanical engineering [44]. Such problem involves seven design variables which are represented in Fig. 9.7. These variables are the face width (x_1), the module of the teeth (x_2), the number of the teeth of pinion (x_3), the length of the first shaft between bearings (x_4), the length of the second shaft between bearings (x_5), the diameter of the first shaft (x_6) and the diameter of the second shaft (x_7).

The objective is to minimize the total weight of the speed reducer considering nine constraints and the physical limits of each variable. Therefore, the mathematical formulation is summarized in Table 9.8.

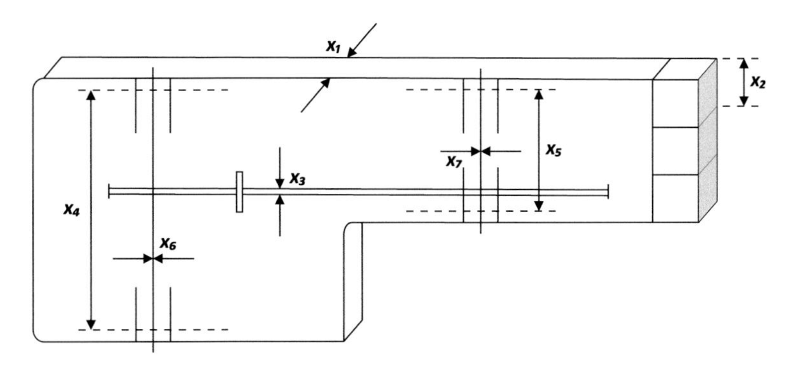

Fig. 9.7 The speed reducer and its design features

Table 9.8 Speed reducer problem

Problem: speed reducer problem	
Minimize	$J_{P3}(\mathbf{x}) = 0.7854x_1x_2^2(3.3333x_3^2 + 14.9334x_3 - 43.0934)$ $- 1.508x_1(x_6^2 + x_7^2) + 7.477(x_6^3 + x_7^3) \cdots + 0.7854(x_4x_6^2 + x_5x_7^2)$
Subject to:	$g_1(\mathbf{x}) = \frac{27}{x_1x_2^2x_3} - 1 \leq 0$
	$g_2(\mathbf{x}) = \frac{397.5}{x_1x_2^2x_3^2} - 1 \leq 0$
	$g_3(\mathbf{x}) = \frac{1.93}{x_2x_3x_4^3x_6^4} - 1 \leq 0$
	$g_4(\mathbf{x}) = \frac{1.93}{x_2x_3x_5^3x_7^4} - 1 \leq 0$
	$g_5(\mathbf{x}) = \frac{\sqrt{\left(\frac{745x_4}{x_2x_3}\right)^2 + 1.69 \times 10^6}}{110x_6^3} - 1 \leq 0$
	$g_6(\mathbf{x}) = \frac{\sqrt{\left(\frac{745x_4}{x_2x_3}\right)^2 + 157.5 \times 10^6}}{85x_7^3} - 1 \leq 0$
	$g_7(\mathbf{x}) = \frac{x_2x_3}{40} - 1 \leq 0$
	$g_8(\mathbf{x}) = \frac{5x_2}{x_1 - 1} - 1 \leq 0$
	$g_9(\mathbf{x}) = \frac{x_1}{12x_2} - 1 \leq 0$
With boundary conditions $2.6 \leq x_1 \leq 3.6$, $0.7 \leq x_2 \leq 0.8$, $17 \leq x_3 \leq 28$, $7.3 \leq x_4 \leq 8.3$, $7.3 \leq x_5 \leq 8.3$, $2.9 \leq x_6 \leq 3.9$ and $5.0 \leq x_7 \leq 5.5$	

Table 9.9 gives the comparison of the SSO-C results with the other methods considering 30 independent executions. Although all methods approximately reach the best value (the optimal solution), the SSO-C algorithm presents the best possible stability (SD). Such fact can be interpreted as the SSO-C capacity of obtaining the best value a higher number of times in comparison with the other algorithms. The best solution obtained by SSO-C is $\mathbf{x}^* =$(3.500000, 0.70000, 17.00000, 7.30001, 7.71532, 3.35021, 5.28665) with $J_{P3}(\mathbf{x}^*) = 2996.11329802963$.

Table 9.9 Statistical features of the results obtained by various algorithms on the speed reducer design optimization problem

	B	M	W	SD
PSO	3044.45297067327	3079.26238870675	3177.51515620881	26.2173114313142
ABC	2996.11571099399	2998.06283682786	3002.75649061430	6.35456225271198
FF	2996.94726112401	3000.00542842129	3005.83626814072	8.35653454754367
SSO-C	2996.11329802963	2996.11329802963	2996.11329802963	1.33518419604E −12

9.5 Summary

In this chapter, a novel swarm algorithm, called the Social Spider Optimization (SSO-C) has been presented for solving constrained optimization tasks. The SSO-C algorithm is based on the simulation of the cooperative behavior of social-spiders. In the presented algorithm, individuals emulate a group of spiders which interact to each other based on the biological laws of the cooperative colony. The algorithm considers two different search agents (spiders): males and females. Depending on gender, each individual is conducted by a set of different evolutionary operators which mimic different cooperative behaviors that are typical in a colony.

For constraint handling, the presented algorithm incorporates the combination of two different paradigms in order to direct the search towards feasible regions of the search space. In particular, it has been added: (1) a penalty function which introduces a tendency term into the original objective function to penalize constraint violations in order to solve a constrained problem as an unconstrained one; (2) a feasibility criterion to bias the generation of new individuals toward feasible regions increasing also their probability of getting better solutions.

In contrast to most of existent swarm algorithms, the presented approach models each individual considering two genders. Such fact allows not only to emulate the cooperative behavior of the colony in a realistic way, but also to incorporate computational mechanisms to avoid critical flaws commonly delivered by the popular PSO and ABC algorithms, such as the premature convergence and the incorrect exploration-exploitation balance.

SSO-C has been experimentally tested considering a suite of 8 benchmark constrained functions and three real-word engineering problems. The performance of the presented approach has been also compared to modified versions of PSO [24], ABC [25] and FF [26] which have been adapted to work over constrained optimization problems. Results have confirmed an acceptable performance of the presented method in terms of the solution quality and stability.

The SSO's remarkable performance is associated with three different reasons: (i) their operators allow a better particle distribution in the search space, increasing the algorithm's ability to find the global optima; (ii) the division of the population into different individual types, provides the use of different rates between exploration and exploitation during the evolution process; and (iii) the constraint handling mechanism allows efficiently to conduct unfeasible solutions to feasible solutions even when new individuals are generated.

References

1. Bonabeau, E., Dorigo, M., Theraulaz, G.: Swarm Intelligence: From Natural to Artificial Systems. Oxford University Press Inc, New York (1999)
2. Kassabalidis, I., El-Sharkawi, M.A., II Marks, R.J., Arabshahi, P., Gray, A.A.: Swarm intelligence for routing in communication networks. In: Global Telecommunications Conference, GLOBECOM '01, 6, pp. 3613–3617. IEEE (2001)
3. Kennedy, J., Eberhart, R.: Particle swarm optimization. In: Proceedings of the 1995 IEEE International Conference on Neural Networks, vol. 4, pp. 1942–1948, Dec 1995
4. Karaboga, D.: An idea based on honey bee swarm for numerical optimization. Technical Report-TR06. Engineering Faculty, Computer Engineering Department, Erciyes University (2005)
5. Passino, K.M.: Biomimicry of bacterial foraging for distributed optimization and control. IEEE Control Syst. Mag. **22**(3), 52–67 (2002)
6. Hossein, A., Hossein-Alavi, A.: Krill herd: a new bio-inspired optimization algorithm. Commun. Nonlinear Sci. Numer. Simulat. **17**, 4831–4845 (2012)
7. Yang, X.S.: Engineering Optimization: An Introduction with Metaheuristic Applications. Wiley, New York (2010)
8. Rajabioun, R.: Cuckoo optimization algorithm. Appl. Soft Comput. **11**, 5508–5518 (2011)
9. Bonabeau, E.: Social insect colonies as complex adaptive systems. Ecosystems **1**, 437–443 (1998)
10. Wang, Y., Li, B., Weise, T., Wang, J., Yuan, B., Tian, Q.: Self-adaptive learning based particle swarm optimization. Inf. Sci. **181**(20), 4515–4538 (2011)
11. Wan-li, X., Mei-qing, A.: An efficient and robust artificial bee colony algorithm for numerical optimization. Comput. Oper. Res. **40**, 1256–1265 (2013)
12. Wang, H., Sun, H., Li, C., Rahnamayan, S., Jeng-shyang, P.: Diversity enhanced particle swarm optimization with neighborhood. Inf. Sci. **223**, 119–135 (2013)
13. Banharnsakun, A., Achalakul, T., Sirinaovakul, B.: The best-so-far selection in artificial bee colony algorithm. Appl. Soft Comput. **11**, 2888–2901 (2011)
14. Gordon, D.: The organization of work in social insect colonies. Complexity **8**(1), 43–46 (2003)
15. Lubin, T.B.: The evolution of sociality in spiders. In: Brockmann, H.J. (ed.) Advances in the Study of Behavior, vol. 37, pp. 83–145 (2007)
16. Uetz, G.W.: Colonial web-building spiders: balancing the costs and benefits of group-living. In: Choe, E.J., Crespi, B. (eds.) The Evolution of Social Behavior in Insects and Arachnids, pp. 458–475. Cambridge University Press, Cambridge, England (1997)
17. Aviles, L.: Sex-ratio bias and possible group selection in the social spider *Anelosimus eximius*. Am. Nat. **128**(1), 1–12 (1986)
18. Burgess, J.W.: Social spacing strategies in spiders. In: Rovner, P.N.: Spider Communication: Mechanisms and Ecological Significance, pp. 317–351. Princeton University Press, Princeton (1982)
19. Maxence, S.: Social organization of the colonial spider Leucauge sp. in the Neotropics: vertical stratification within colonies. J. Arachnol. **38**, 446–451 (2010)
20. Eric, C., Yip, K.S.: Cooperative capture of large prey solves scaling challenge faced by spider societies. In: Proceedings of the National Academy of Sciences of the United States of America, vol. 105(33), pp. 11818–11822 (2008)

21. Michalewicz, Z., Schoenauer, M.: Evolutionary algorithms for constrained parameter optimization problems. Evol. Comput. **4**(1), 1–32 (1995)
22. Michalewicz, Z., Deb, K., Schmidt, M., Stidsen, T.: Evolutionary algorithms for engineering applications. In: Miettinen, K., Neittaanmäki, P., Mäkelä, M.M., Périaux, J. (eds.) Evolutionary Algorithms in Engineering and Computer Science, pp. 73–94. Wiley, Chichester (1999)
23. Koziel, S., Michalewicz, Z.: Evolutionary algorithms, homomorphous mappings, and constrained parameter optimization. Evol. Comput. **7**(1), 19–44 (1999)
24. He, Q., Wang, L.: A hybrid particle swarm optimization with a feasibility-based rule for constrained optimization. Appl. Math. Comput. **186**, 1407–1422 (2007)
25. Karaboga, D., Akay, B.: A modified Artificial Bee Colony (ABC) algorithm for constrained optimization problems. Appl. Soft Comput. **11**, 3021–3031 (2011)
26. Amir, G., Xin-She, Y., Amir, A.: Mixed variable structural optimization using firefly algorithm. Comput. Struct. **89**, 2325–2336 (2011)
27. Gan, M., Peng, H., Peng, X., Chen, X., Inoussa, G.: An adaptive decision maker for constrained evolutionary optimization. Appl. Math. Comput. **215**, 4172–4184 (2010)
28. Oster, G., Wilson, E.: Caste and ecology in the social insects. N.J. Princeton University Press, Princeton (1978)
29. Hölldobler, B., Wilson, E.O.: Journey to the Ants: A Story of Scientific Exploration (1994). ISBN 0-674-48525-4
30. Hölldobler, B., Wilson, E.O.: The Ants. Harvard University Press, Cambridge (1990). ISBN 0-674-04075-9
31. Avilés, L.: Causes and consequences of cooperation and permanent-sociality in spiders. In: Choe, B.C. (ed.) The Evolution of Social Behavior in Insects and Arachnids, pp. 476–498. Cambridge University Press, Cambridge (1997)
32. Rayor, E.C.: Do social spiders cooperate in predator defense and foraging without a web? Behav. Ecol. Sociobiol. **65**(10), 1935–1945 (2011)
33. Gove, R., Hayworth, M., Chhetri, M., Rueppell, O.: Division of labour and social insect colony performance in relation to task and mating number under two alternative response threshold models. Insect. Soc. **56**(3), 19–331 (2009)
34. Ann, L., Rypstra, R.S.: Prey size, prey perishability and group foraging in a social spider. Oecologia **86**(1), 25–30 (1991)
35. Pasquet, A.: Cooperation and prey capture efficiency in a social spider, *Anelosimus eximius* (Araneae, Theridiidae). Ethology **90**, 121–133 (1991)
36. Ulbrich, K., Henschel, J.: Intraspecific competition in a social spider. Ecol. Model. **115**(2–3), 243–251 (1999)
37. Jones, T., Riechert, S.: Patterns of reproductive success associated with social structure and microclimate in a spider system. Anim. Behav. **76**(6), 2011–2019 (2008)
38. Damian, O., Andrade, M., Kasumovic, M.: Dynamic population structure and the evolution of spider mating systems. Adv. Insect Physiol. **41**, 65–114 (2011)
39. Yang, X.-S.: Nature-Inspired Metaheuristic Algorithms. Luniver Press, Beckington (2008)
40. Chen, D.B., Zhao, C.X.: Particle swarm optimization with adaptive population size and its application. Appl. Soft Comput. **9**(1), 39–48 (2009)
41. Liang, J.J., Runarsson, T.P., Mezura-Montes, E., Clerc, M., Suganthan1, P.N., Coello, C.A.C., et al.: Problem definitions and evaluation criteria for the CEC 2006 special session on constrained real-parameter optimization. Technical Report #2006005. Nanyang Technological University, Singapore (2005)
42. Coello, C.A.C.: Use of a self-adaptive penalty approach for engineering optimization problems. Comput. Ind. **41**, 113–127 (2000)
43. Cagnina, L.C., Esquivel, S.C., Coello, C.A.C.: Solving engineering optimization problems with the simple constrained particle swarm optimizer. Informatica **32**, 319–326 (2008)
44. Jaberipour, M., Khorram, E.: Two improved harmony search algorithms for solving engineering optimization problems. Commun. Nonlinear Sci. Numer. Simulat. **15**, 3316–3331 (2010)